Translational Systems Sciences

Volume 27

Kyoichi Kijima • Junichi Iijima • Ryo Sato •
Hiroshi Deguchi • Bumpei Nakano
Editors

Systems Research II

Essays in Honor of Yasuhiko Takahara on Systems Management Theory and Practice

Springer

Editors
Kyoichi Kijima
School of Business Management
Bandung Institute of Technology
Kawasaki, Japan

Junichi Iijima
School of Management
Tokyo University of Science
Tokyo, Japan

Ryo Sato
School of Management
Tokyo University of Science
Tokyo, Japan

Hiroshi Deguchi
Faculty of Commerce and Economics
Chiba University of Commerce
Tokyo, Japan

Bumpei Nakano
Yokohama, Japan

ISSN 2197-8832 ISSN 2197-8840 (electronic)
Translational Systems Sciences
ISBN 978-981-16-9940-5 ISBN 978-981-16-9941-2 (eBook)
https://doi.org/10.1007/978-981-16-9941-2

This Springer imprint is published by the registered company Springer Nature Singapore Pte Ltd.
The registered company address is: 152 Beach Road, #21-01/04 Gateway East, Singapore 189721, Singapore

Preface

It is a great pleasure to publish Systems Research I and II: Essays in Honor of Yasuhiko Takahara, to commemorate the 50th anniversary of Dr. Yasuhiko Takahara's research and education activities, who has been active at the world level in the field of systems research. We compile representative research of researchers and practitioners scented by Dr. Takahara from Japan and abroad who pay homage and gratitude to him into two volumes.

Dr. Takahara completed his research with Dr. M. Mesarovic at Case Western Reserve University. He brought up the results of General Systems Theory (GST), especially Mathematical General Systems Theory (MGST), from the United States to Japan to be appointed to Tokyo Institute of Technology in 1972. In Japan, it was a time the term "system engineering" began to attract interest in gradually establishing the system as a unique field and capturing the essence of what is recognized as a system.

We, editors-in-chief, who fortunately shared that era, still remember the shock we had when we first learned GST, especially MGST. MGST attempts to transparently understand the properties of systems such as interdependency and emergent property, by discussing the logic related to systems in a set-theoretic framework to formulate causal systems and hierarchical systems.

A wide range of books used as textbooks at the seminars in Takahara laboratory remind us of such good old days. They not only have worked as the soils supporting our research since then but also certainly reflect some part of the background of knowledge at that time. They include: *Universal Algebra* (George Graetzer); *Algebra* (Saunders MacLane and Garrett Birkoff); *Introduction to Topology and Modern Analysis* (George F. Simmons); *Topology* (James Dugundji); *Abstract and Concrete Categories: The Joy of Cats* (Jiri Adámek, Horst Herrlich, and George E. Strecker); *Model Theory* (C.C. Chang, H. Jerome Keisler); *Beginning Model Theory* (Jane Bridge); *Model-Based Systems Engineering* (A. Wayne Wymore); *Theories of Abstract Automata* (Michael A. Arbib); *The Specification of Complex Systems* (B. Cohen et al.); *The Structure of Scientific Theories* (Frederick Suppe (Editor)); *Goedel, Escher, Bach* (Douglas R. Hofstadter); *Forever Undecided* (Raymond Smullyan); *Introduction to Systems Philosophy* (Ervin Laszlo); *Living Systems*

(James Grier Miller); *Facets of Systems Science* (George J. Klir); *Systems Thinking, Systems Practice* (Peter Checkland); *Heuristics* (Judea Pearl); *Multifaceted Modelling and Discrete Event Simulation* (Bernard P. Zeigler).

Over the 50 years since then, Dr. Takahara has developed the concept and theory of general systems centered on formal system research. By projecting mathematical general systems theory from meta-theory to the real world, he has powerfully promoted diverse but coherent systems research with a strong desire and intention to construct knowledge not only in theory (episteme) but also in engineering (techne) and practice (phronesis). They include systems modeling, information systems, decision support systems, and systems thinking. The results and findings there are now having a great impact on improving our socio-economic situations, which are becoming more and more complex, by providing the basis of ideas for the sophistication of business models and creation of new services by networks and ICT.

At the same time, Dr. Takahara has promoted educational activities with these studies and research as a nursery to give a great influence on many researchers and practitioners, not limited to students who received his direct scent. It triggers intellectual excitement and acts as a device to encourage their diversified but coherent systems study.

The Takahara School of GST is outstanding in its interdisciplinary and transdisciplinary approach. As you can see in the table of contents of the book, it ranges from highly abstract mathematics to practical applications of social science. It is very fortunate in systems research history that the Takahara school has created such a wide range of content in a deep dialogue using "system" as the common concept.

Some 20 authors gathered here have established their position by developing systems concepts in theory, models, methodologies, and applications in various ways, keeping in their mind the works by Dr. Takahara and co-authored with Mesarovic. Their activities include:

- Research to develop the strong intellectual desire for generality tackled by mathematical general systems theory into knowledge to connect different levels and approaches to the same object for solving actual problems. They are eager to catch the spirit that GST aimed at in the early days.
- Enterprise by which MGST and system engineering are practically fused by examining the basic concept of the system to develop knowledge applicable to hot topics these days.
- Development of what we call translational approach that connects theory and practice in a cyclic way, which follows the process of analyzing mechanisms, solving in an evidence-based way, and intervening in a problematic situation.
- Exploration of new disciplines such as Decision Systems Science and Service Systems Science by accommodating "hard" systems theory with "soft" systems thinking.
- Promotion of a formal approach in the field of Information Systems, for example, to identify isomorphisms between different recommendation systems

and decision schemes known in social choice theory, and to the Enterprise Ontology in Design and Engineering Methodology for Organization.

Since the articles contributed by the 20 authors have a wide range of issues, we decided to structure them based on topics and approaches and publish them as a two-volume set. The second volume consists of 13 chapters divided into 2 parts dealing with the field of Systems Management Theory and Practice.

Finally, we would like to express deep thanks to you Yutaka Hirachi of Springer Japan and Selvakumar Rajendran, and other staff of SPi Global. Without their patience and understanding, this volume could not be published.

Kawasaki, Japan — Kyoichi Kijima
Tokyo, Japan — Junichi Iijima
Tokyo, Japan — Ryo Sato
Tokyo, Japan — Hiroshi Deguchi
Yokohama, Japan — Bumpei Nakano
December 2021

Contents

Part I
Social/Organizational Theory and Application

Chapter 1
Complex Systems and Postmodernism: A New Perspective for Society in the Twenty-First Century

Takatoshi Imada

Abstract The objective of this chapter is to claim that the Hobbesian problem asking how the social order is possible must be replaced with the problem of how the edge of chaos is established. Fluctuation, noise, disequilibrium, and chaos have been noticed and emphasized in the last two decades as the major factors in the science of complexity. We may say that the current of the new scientific view of complex systems indicates the passing of a *Copernican Change* from the order theory to the chaos theory of the world in the twenty-first century. The postmodernism succeeding the poststructuralism is a social version of the science of complexity and offers the viewpoint that foresees the self-organization of the coming new civilization. In the 'edge of chaos' in modern civilization, the social system enters an excitation state and the 'rhizomic (schizophrenic)' movement dedifferentiates the modern structural-functional system. Social roles and statuses formed with the functional requirement are dedifferentiated, and the actions not constrained by them become dominant. In addition, control and manipulation become invalidated and concepts such as control center, authorities and wholeness of system are refuted. The increase of the arguments regarding complex systems and postmodernism symbolically signals that modern civilization has reached the edge of chaos. Probably, a plan of the new civilization will come out from the chaos theory of the world. In the twenty-first century, the situation of society seems to have become heated by the generation of noise, fluctuation, and chaos.

Keywords Edge of chaos · Self-organization · Rhizomic system · Dedifferentiation · Lack of transparency of representation

---

This chapter is a translation of the significantly revised version of my lecture record (Imada, 2001b).

T. Imada (✉)
Emeritus Professor at Tokyo Institute of Technology, Tokyo, Japan

K. Kijima et al. (eds.), *Systems Research II*, Translational Systems Sciences 27,
https://doi.org/10.1007/978-981-16-9941-2_1

When social science was evolving as a field along with the rise of modern society, Thomas Hobbes presented a famous problem that he had raised in his book *Leviathan*—how will social order be possible once the notion of God disappears from society? While in Medieval Ages, social order was maintained through the concept of the metaphysical existence of God, the dawn of the Modern Age witnessed "the God killing" through a series of movements such as the Renaissance, Reformation, scientific revolution, and the trial that established human control over the world. However, social order was not formed and maintained easily without God, because human beings, when released from God's control, begin to selfishly pursue personal profit pursuit, thereby resulting in a state of "the war of everyone against everyone," which will continue if the situation prevails. Therefore, the problem of how a society and order can be formed and sustained without premising the existence of God needs to be solved. The quest for the solution to this Hobbesian problem of order led to the creation of the academic field that is called social science today.

The above Hobbesian problem has been considered by many prominent thinkers such as John Locke, Montesquieu and Rousseau; finally, it was Adam Smith who suggested a provisional solution to the problem by formulating the market mechanism as the "God's invisible hand." Since then, the process of formation of a society assuming freedom and equality for all individuals has been refined. To put it simply and precisely, one can say that the essence of modern social science involves a struggle with the problems related to order that arise after the disappearance of God.

However, the present situation requires rethinking of the current scientific view. We come across a new reality that cannot be grasped by modern science. The necessity of noticing the rich possibility of the existence of this reality in the excitation state at the boundary between order and chaos is evident. Thus, what we have is a movement intending to discover the frontier of a new science wherein there exists a reality that is neither chaos nor order, and also both chaos and order. Such a trend indicates that the Hobbesian problem of order must be replaced with the problem of how the edge of chaos is established.

1.1 From Order Theory to Chaos Theory of the World

Although modern society has been supported by the thought of progress and growth, the focus has shifted toward the problem of chaos rather than that of order since the last quarter of the twentieth century. Further, phenomena such as fluctuation, noise, disequilibrium, and complexity are interestingly being emphasized. In particular, in the past 40 years, an increasing amount of attention is being given to the dynamic movement—which is occurring in the boundary between chaos and order—rather than the problem of how the formation and sustainability of order and equilibrium. Although, modern science focuses on the formation and maintenance of order, new currents focus on the order problem reversely from the viewpoint of "edge of chaos" that is a transition zone between order and disorder.

The above is my real argument, which can also be summarized as follows: the essence of the science of complex systems lies in replacing the order problem with the problem of the establishment of the "edge of chaos." My contention is that the implication of the science of complex systems lies in the shift "from the order theory to the chaos theory of the world."

1.1.1 A Copernican Revolution

The paradigm revolution of science that has advanced since the middle of the 1980' is considered to be more dramatic, as the shift is equal to the Copernican Revolution from the geocentric theory to heliocentric theory. The heliocentric theory directed the scientific revolution that was passed down from Galileo to Newton, and it became the thesis that symbolizes the modern science. Drawing an analogy, we can state that the current trends seen in the new scientific view emerging at the turn of the twentieth century in the areas concerning the self-organization theory, such as fluctuation, noise, and chaos, can be interpreted as an indication of the Copernican Revolution from the order theory to the chaos theory.

Although the objection that the chaos theory of the world is a reckless expression is anticipated, the meaning of the term "chaos" here is different from that in ordinary language. When we refer to chaos in ordinary life, we imply that the distinction between things is unclear, and there is lack of order. Hence, chaos is not used in this ordinary sense but is scientifically defined as the condition wherein even if we are in a position to interpret the present state of affairs it is impossible to predict the future because small changes in initial conditions produces large changes in long-term outcome (so-called butterfly effect), that is "sensitivity to initial conditions" (Lorenz, 1963; Gleick, 1987: Chap.1).

The most important aspect of the chaos theory is grasping what occurs at the edge of chaos as found in the science of chaos. Otherwise, the chaos theory does not hold and the decoding of the world is inaccurate.

In modern science, since the time of Galileo and Newton, reductionism, determinism, reversibility of time, equilibrium, and stability have been emphasized. Thus, it is the responsibility of science to extract law (≒ order) from the world. There exists a background hypothesis that it is possible to articulate order as something that is separate and distinct from chaos.

However, in the self-organization paradigm in a broad sense that was consciously and widely conceptualized after the late 1970s, emergent property, nondeterminism, irreversibility of time, nonequilibrium, instability, different from Galileo-Newtonian science are emphasized (Nicolis & Prigogine, 1977; Haken, 1976; Jantsch, 1980; Imada, 1986). This paradigm has a background hypothesis that apparently paradoxical conditions such as order with chaos or an interpenetration between order and chaos is the essence of the world. In short, the paradigm includes a scientific view that the world must be an unsteady system that fluctuates at the boundary between order and chaos. Chaos and order cannot be clearly distinguished and the system has a boundary that delicately fluctuates between the two.

This new change in the scientific view also encompasses the claim that order cannot exist without chaos and the edge for interpenetrate with chaos. This, I assume, is the chaos theory of the world; further, I assume that this aims to deconstruct the concept of the modern age that the discovery and realization of order are important.

While we find such a scientific trend only in the fluctuation theory that was popular in the latter half of the 1970s, the theory of noise and chaos appeared in the 1980s. This was followed by the theory of complex systems—which included all of the abovementioned views—that was advocated in the 1990s.

The chaos theory does not exclude order but asserts that the "edge of chaos" is of decisive importance. Therefore, it does not recognize the significance of random noise and fluctuation that only disturb or destroy order. The chaos theory emphasizes noise and fluctuation that have the properties of both emergence and amplification. Therefore, the chaos theory can create a common front with the paradigm of classical modern science in the sense that it controls noise and fluctuation that are not located at the edge of chaos. However, the chaos theory diverges from modern science in that it does not seek to make fluctuation and noise that are located at the edge of chaos objects of control.

Fluctuation, noise, and chaos have been considered "outlaws" by modern science. These are located outside order and are considered to be factors that disturb order. In the last four decades, however, these were spotlighted as the source of spontaneous order, comfort, and factors that enable self-organization. In my view, the science of complex systems has been advocated as the field where all fluctuations, noise, and chaos converge.

1.1.2 Edge of Chaos and Self-Organization

The essential point of the science of complex systems is not to emphasize chaos excessively but to focus on the edge of chaos (Packard, 1988; Langton, 1990). At the edge of chaos, the phase transition of the system frequently occurs. Further, the emergence of new information is activated without dissolving the order. However, the system becomes unstable and the fluctuation is frequently generated. The system enters an excitation state and what happens is unknown. Under these conditions, mutations, and information emergence occur frequently.

Systems reached at the edge of chaos display characteristics of Kaneko and Ikegami's (1998, Chap. 3) "homeochaos." This compound concept of homeostasis and chaos is a type of chaotic state that appears in the ecosystem model of parasite and host. Although the ecosystem becomes unstable when the parasitism for the host heightens, this chaotically unstable system can be dynamically stabilized. Such stabilization is possible in the case of the development of a model in which a system can change the mutation rate on its own by diversifying the species through an increase in the mutation rate. Homeochaos is the chaotic state with such a degree of freedom. In other words, it is homeostasis that includes the chaos.

Homeochaos is that state of a system that includes fluctuation (mutation) at the edge of chaos. The edge of chaos is the place where the system is in an excitation state, new information emerges, variation actively arises, and fluctuation occurs frequently. In this sense, we can call homeochaos a higher order homeostasis that incorporates morphogenesis.

The above is an outline of chaos theory of the world that the new system paradigm has brought about. Chaos theory does not entrust the fate of the world to randomness. While the Galileo-Newtonian modern science prioritizes order, the chaos theory is significant in that it advances the elucidation of the complexity by giving the appropriate value to chaos.

1.2 Science of Complex Systems: Convergence of Various Chaos Theories

As previously mentioned, the science of complex systems has been advocated as the field where all the fluctuations, noise, and chaos that are positioned as outlaws of modern science converge. While the term "complex" usually implies a condition wherein we cannot understand things or situations clearly due to their complicatedness, the science of complex systems intends to scientifically deal with the system in such a condition. In this sense, complex systems can refer to chaos in the ordinary language. In academic usage, however, the term "complex systems" does not merely imply a complicated condition wherein large numbers of elements interact irregularly. It is the essence of a complex system, even if it consists of a small number of elements, to immanently remain unstable, thereby precluding the prediction of its behavior.

Sociology also developed "complexity" into a theme. For example, sociologist Luhmann (1972, 1984) considers "reduction of the complexity" as one of the key concepts in social system theory. However, complexity in the reduction of complexity differs from the complexity in the science of complex systems, which is our topic of interest at present.

Luhmann's "reduction of complexity" is a paraphrase of control theory in cybernetics, especially Ashby's (1956, Chaps. 7 and 11) "law of requisite variety." According to this law, a system must have an equal number or more varieties (number of effective strategy) than the environment for the system to achieve its purpose through effective control of the environment. In this sense, the law of requisite variety is a principle of control theory for goal attainment. Luhmann's contrivance is that the system not only maintains order through the reduction of complexity but also preserves complexity in the level of meaning. However, while it is important to assess the behavior of semantically pooled complexity (variety), Luhmann does not give priority to this aspect in his research.

The problem of complexity was also the main theme when the general system movement reached its peak in the latter half of the 1950s. While the general system theory proposed the unification of specialized disciplines, one of its claims was the unification of organicism and mechanism (von Bertalanffy, 1968). By that time, organicism was referred to as the theory clarifying "chaotic complexity" and mechanism was referred to as the theory clarifying "organized simplicity." Similarly, the general system theory was positioned as unified science for dealing with "organized complexity" in order to overcome this undesirable condition.

As compared to the present science of complex systems, the context of the problem of complexity in the general system movement is different. The aim of the general system movement was, as just mentioned, to establish a discipline that would deal with organized complexity by overcoming the undesirable antagonism between organicism and mechanism.

The general system theory had no viewpoint that consciously based the chaos theory of the world on the problem of the complexity. It just tries to incorporate the organismic properties such as wholeness, order, goal directionality, growth, and differentiation in mechanism. Regarding this point, the general system theory may be distinguished from the science of complex systems (which are more up-to-date).

In the background of the science of complex systems, there is skepticism that reality cannot be accurately recognized from the methodological standpoint, which modern science considers to be a golden rule. The first issue pertains to reductionism and questions whether reductionism is sufficient in order to recognize a phenomenon. The second argument is about the non-recognition of cognitive processes and ability of order formation in the object.

There is no new criticism of reductionism because the claim that "the wholeness is over the sum of the parts" is merely an extension of conventional organicism and general system theory. It has been asserted for a long time that an emergent property, which cannot be reduced to its elements, will be generated by the interaction among elements. A new line of thinking, however, does exist in the second argument. In a social phenomenon, it is reasonable fact to recognize the cognitive process and accept the ability of the order formation in the object; however, these characteristics often exceed the recognition range of conventional scientific methods, wherein the methodology of natural science enjoys superiority. Therefore, we will simply use the metaphor "let the world tell itself."

Complex systems always have the dynamism of an unanticipated change since many elements irregularly interact in various ways. Alternatively, it is possible for the elements to self-organize and generate a new structure and order through fluctuation. We can observe the characteristics of complex systems through the following examples: the irregular changes seen in stock prices and exchange rates, meteorological phenomena that are difficult to predict, life generated from complicated molecular combinations, the manner in which the brain produces feeling and thought.

Another reason that the research of complex systems has become so popular is the rapid development of computational science. Complex systems generally comprise interactions between large numbers of elements, and it is impossible to solve these by analytical mathematical modeling. With significant increases in computational capacity, however, it becomes possible to simulate a system (even those that cannot be solved through mathematical modeling) and predict its behavior.

In modern computer science, the conversion from procedure-oriented approaches to the object-oriented ones is now fairly advanced; further, the method used for assigning some function to data and variables has been developed. In conventional simulation, it was a premise that an object behaves according to the procedure (the rule) decided beforehand. Object-oriented simulation, however, recognizes the cognitive process of the object and its ability to decide its behavior on its own. That is, by receiving only a small number of constraints, objects such as variables and data autonomously interact in the computer and can form the whole system. By replacing this data and variables with human (agent) behavior, we can expect to obtain the consequences of this interaction, which cannot be obtained analytically.

A well-known example is the "Tit for Tat" strategy obtained from the repetition prisoner's dilemma game. The prisoner's dilemma game is the game that consists of two strategies: cooperate (stay silent) and defect (betray). However, it is impossible to analytically identify the type of strategy that is effective when we repeat this game. According to the result of a simulation, the most effective strategy in a repetitive prisoner's dilemma game is Tit for Tat—first cooperate and then adopt the strategy that the partner has used in the previous game. Axelrod (1984) called this strategy the "evolutionary stable strategy"; further, attempts to discover this solution analytically failed. Thus, the object-oriented (agent-based) simulation has become a powerful method for clarifying complex systems.

The remarkable feature of the science of complex systems is that on applying computer simulation, it leads to unexpected results regarding system behavior, which are not obtained analytically. This is an important issue as it limits the applicability of analytical methods based on the Galileo-Newtonian type of reductionism.

The summary of Waldrop (1992) is accurate in that complex systems have the capability to form a special equilibrium that includes both order and chaos. The systems built at the edge of chaos are complex systems. Complex systems are located neither in chaos nor in order, and demonstrate the self-organizing property in the "battlefield between stagnation and anarchy."

In sum, the characteristics of the science of complex systems are as follows: (1) having the object-oriented perspective that recognizes the cognitive process of the object, (2) emphasizing the "edge of chaos" at the boundary between order and disorder, and (3) investigating the condition for enabling the "self-organization" of the system. These characteristics show that the science of complex systems plays an important role when it comes to the conversion from order theory to chaos theory.

1.3 Postmodernism or "Edge of Chaos" in Civilization

Postmodernism is a phenomenon at the edge of the chaos of modern civilization. That is, postmodernism is the phenomenon showing that modern civilization is located at the edge of chaos, which leads to the phase transition (metamorphosis) to a different civilization.

I regard the thought of poststructuralism and postmodernism (which immediately followed) as a social version of the change from the order theory to chaos theory in the paradigm of science. For the concept of social order, which the modern civilization has aimed for, these paradigms try to investigate the condition of the self-organization of a new civilization by introducing the viewpoint of the chaos. In poststructuralism (Deleuze & Guattari, 1980), the chaos theory of the world typically appears in the concept of "difference" and "rhizome" (subterranean stems such as lotus and bulb) and in postmodernism (Poster, 1990; Lash, 1990), in the "lack of transparency of the representation" and "dedifferentiation." For the modern civilization, these are the phenomena located at the edge of chaos.

1.3.1 Difference and Rhizome

Poststructuralism asserts the objection against the structural concept raised by structuralism. It moved to the forefront as a reaction against the notion that structuralism as represented by Levi-Strauss (1958) excluded such phenomena as violence, noise, differences, etc. from the structural concept. The typical definition of a structure in structuralism is that it is a system of transformation rules. According to this definition, a structure, as a system of ordered rules, becomes a premise; however, poststructuralism claims that a structure is not strict but has various cracks, frictions, and gashes. It is difference and heterogeneity that bring about these cracks. Therefore, in poststructuralism, the dynamics of the difference are pivotal.

Differencification (I will use this term to refer to the dynamics of differences in poststructuralism, which are different from functional differentiation) is the movement that intermingles the heterogeneity and brings about further cracks and gashes in the structure. Poststructuralism does not emphasize only differencification or structural cracks and gashes. In the method of deconstruction, reconstruction work after the structural disassembling is emphasized. As Derrida (1972) puts it, deconstruction does not mean mere destruction or denial but also includes the reconstruction of the structure, if necessary. While Derrida makes limited use of the method of deconstruction to study philosophical and literary discourse, it is applicable to social system theory.

While framing such an approach, the rhizomic system, consisting of the phantasmagoric movement of differences, is assumed. Although the concept of "rhizome" was devised by Deleuze and Guattari (1980), the term actually refers to subterranean stems such as a lotus and bulb. Rhizome is not a root but a tree that grows underground. It is an example of differencification and is not considered to be in order, much like the relation between a tree and root. Thus, the rhizomic system does not have a center and rank ladder. It is a manifold wherein there is no distinction between order and chaos and between subjects and objects. Rhizome is a movement body that deconstructs the modern concepts of order and equilibrium, and an anti-control system that accommodates fluctuation.

Rhizome is always incomplete and is multi-dimensional manifold that lacks one dimension. For this reason, new differencification will arise. Hence, the rhizome is always transforming to a different manifold. In addition, it does not have a hierarchy structure and there is no center either. Rhizome is a chaotic system that includes the interaction in which various elements are entangled. Moreover, the rhizome's arbitrary node can be connected to other arbitrary nodes. Events such as cutting, decomposition, assembly, connection, and turning over are possible in all dimensions and there always exists the possibility of the acceptance of change. I regard this property as the "principle of free coupling," and rhizome displays phantasmagoric movement through this principle (Imada, 1987, 1994).

The hypertext world of electronic media displays this property. The retrieval method of a conventional database is of the tree type. In the hypertext world, however, it is possible to fly or warp from one arbitrary text to another arbitrary one. The system is free of locational constraints. Therefore arbitrariness (chance) is permitted. However, as long as the retrieval is incomplete, one is required to go somewhere (necessity). In this sense, the behavior of the hypertext world exhibits the "principle of necessity-chance" (Kuroishi, 1991, Chap. 3). In this case, the World Wide Web is a representative example.

An interesting feature of the rhizome is that it has the logic of self-reference (Imada, 1987). In the formal logic that forms the basis of modern science, self-reference is forbidden because this leads to a paradox. Rhizome is the manifold wherein there is no distinction between subject and object and dualism is inapplicable. Moreover, there is neither the existential dualism of "here and there," nor the value dualism of "good and bad." Rhizome has no boundary separating the inside from the outside because it refuses dualism. Therefore, the paradox of self-reference does not occur in rhizome. It has only limitations pertaining to size and dimensions and lacks a measurable unity. It cannot be assumed that rhizome changes itself while adapting to the environment. That is, rhizome exhibits the "principle of spontaneous order" that allows it to change both its properties and itself on its own.

Along with performative action, the notion of spontaneous order is a central concept of von Hayek's (1973) social philosophy. Performative action is not premeditated and rational action but tacit action from which spontaneous order is formed. While I have considered this type of order formation to be a type of self-organization, such dynamism is rhizomic. The language system has a similar dynamism.

In the daily use of language, every individual is not conscious of grammar (rule) and no one changes the language intentionally. However, thus far, new languages have been produced and grammars have been changed through daily conversation and writing. These changes have been brought about by the rhizomic activities—the differential differencification of language and its usage.

Another characteristic of the rhizome that needs to be emphasized is that it is an anti-control system that does not accept administrative supervision and manipulation (Imada, 1997). The essence of rhizome lies in the "principle of control destruction" that invalidates and destroys the control mechanism. The control mechanism always exists as the object for decomposition and modification. It goes without saying that the control mechanism does not simply collapse and the sanction will be given at the time of destructive action. In order to cope with this, however, rhizome has 'various escape lines', and it can always escape from any place. Although the runaway or escape has excessively been emphasized as one of rhizome's properties until now, it is necessary to recognize control destruction as the important point.

1.3.2 Rhizome as a Principle of Social Formation

While humankind has come up with multiple principles of social formation until now, rhizome will be established as a new principle in the twenty-first century. We have already had three principles of social formation. They are the principles of market, hierarchy, and network. Rhizome is the fourth principle and cannot be reduced to the above three. Market, hierarchy, and network are the social formation principles of the order phase that are constructed by the modern society. On the other hand, rhizome is the social apparatus that generates fluctuation in the phase where both chaos and order exist or in the place where the order phase is collapsing.

Although network is recently being spotlighted in relation to the computerization and informationalization of society, this is the principle that is the characteristic of the modern functional civilization. This is because the concept of function did not originally include a vertical ladder like the hierarchy principle. The essential point of the functional formation is to accomplish work through the role division of labor, and therefore does not require in principle the hierarchical ladder of positions. However, as the method of integration using a network is yet to be fully developed, it has been substituted by hierarchical authority and command line. In addition, as there are differences among the many roles with regard to its performance, even if the network only caters to the functional (horizontal) role division of labor, hierarchical structure of status has been specifically designed to reward achievement.

Improvement in the method of social formation by the network principle will contribute toward resolving some of the problems inherent in the hierarchy principle. However, the problem in question is that the network principle is the modern version of the social integration, especially functional integration. Society always consists of dynamism with integration and disintegration; perfect integration does not exist.

As the network primarily focuses on integration, it is bound to contribute to social integration that allows a greater level of freedom and autonomy than that possible in conventional society. Social dynamism always includes a disrupting factor against the normal factor with regard to integration is always included; thus, it serves as a source of social vitality.

It is true that the network recognizes the diverse connection of the elements, and emphasizes not the vertical connection by the bureaucratic system but the horizontal connection. However, the notion that unity will be sustained is a very major presumption. Rhizome denies this. It often twines around the node of the network, cuts the existing link (connection), and then gives the link to another node and/or extends in a different direction.

In my view, the rhizomic activity is a dynamic phase located at the edge of chaos in modern civilization. While modern civilization has made function primacy social apparatus that emphasizes efficiency and rationality, rhizome deconstructs the logic of the function. Further, it also crushes the structuralist system of transformation rules as an unconscious spiritual structure. Thus, rhizome is exactly transfunctional and nonstructural system (Imada, 2016).

1.3.3 Lack of the Transparency of the Representation

The postmodernism is the general term used to refer to the situation that has arisen after the Modern Age and for the realities after the age which modernism has lost the ability to grasp the reality. The roots of postmodernism lie in the nihilism against the modernist thought of enlightenment and progress. Postmodernism is an extension of the nihilism where there is an unstable and groundless sense for reality.

Postmodernism originated from the criticism of modern architecture, which pursued efficiency and rationality in place of vivid decorations and futility. While the modern architecture intends to investigate a function, postmodernism tries to introduce play, symbolic expressions, free ideas, etc. into architecture. The context for the rise of postmodernism indicates the following: formation of a global economy dominated by multinational enterprises, change from a mainly production society to a consumer one, increase in the symbolic meaning of the media and sings, etc. Pastiche and schizophrenia are also specificities of the postmodern experience.

Postmodernism denies the efficient unity of a system and integration based on a consensus. Its foundations are minor, local, and heterogeneous work and paralogy (anti-logical imagination) of differencification without mutual commensurability. Lyotard (1979), one of the founders of postmodernist thought asserts that we have already reached the age that indicates the end of modern "grand narratives." The present state raises challenges for modern grand narratives with regard to the following aspects: realization of freedom and equality, search for consensus and legitimacy, and aiming toward revolution and emancipation.

The common characteristic of postmodern thought in the fields of social sciences, humanities, philosophy, design, or architecture is the criticism for modern functionalist reason. "Form (structure) follows function" by Louis Sullivan and "beauty is a promise of the function" by Carma Gorman is the essential feature of modern architecture. The essence of modern ages is that the function has colonialized and dominated the structure (form) and the meaning (beauty). Postmodernism intends to overturn this.

Postmodernism tries to crush the transparency of the representation of real existence. The "metaphysics of the presence" in the Modern Age that regards the direct correspondence between language and real existence as self-evident is very doubtful. Postmodernism claims that it is impossible to univocally decide the meaning of a phenomenon and ultimately all matters including science are groundless.

One of the reasons for the crack in the modern image of reality is the loss of meaning owing to the collapse of the direct correspondence between real existence and representation caused by the electronic media, especially virtual reality. When the reliable existence of representation is not guaranteed, reality becomes a substance-lacking fragment of a sign (in other words, a floating bubble). This corresponds to the lack of transparency of representation in postmodernism. The lack of transparency implies that the representation does not have a real world counterpart, that is, a condition of déraciné.

In the modern thought, it is important that representation ensures transparency. (Logical) positivism asserted its importance and neglected representation without transparency as mere idea or fancy. However, postmodernism emphasizes the significance of the representation without transparency and intends to crush transparency. When this becomes extreme, however, we cannot univocally decide the meaning of the phenomenon.

In fact, postmodernism often regards all matters including science as groundless. Although this is an extreme claim, when we consider the rise of virtual reality (owing to the advances made by IT), which is soon going to be socially recognized as another reality, it becomes difficult to make the transparency of the representation a reason for reality (Poster, 1990). In the world of virtual reality, correspondence to existence is not a problem. Therefore, entering the world of virtual reality is equivalent to experiencing the postmodern world, wherein representation does not correspond to existence. An embodiment of a concept that has no correspondence with existence can become a reality very well.

The above discussion destabilizes modernity. Realities created through computer simulation (simulacra) have the same degree of reality as reality. Further, the world of virtual reality, which embodies concepts through stored signs and information in the computer (network), buries us in technology and leads to a paradox where we consider virtual reality to be real despite its fictitious nature. In the world, when activities lack the transparency of representation, the basis of the existence confirmation of the self would be unsteady or people would lose the place where their identities rest and experience a state of homelessness in terms of identity. This is the lack of the transparency of existence corresponding to representation.

That the postmodern society is a consumer society also brings about cracks and losses in the image of reality. In a consumer society, distinctions and boundaries between things blur and thus, there is a lack of transparency of representation. Although a consumer society is one wherein the influence that the consumption mode gives to culture and society increases, the society's characteristics lie in that people buy not only the thing itself but also the symbolic image (meaning) associated with that thing. In symbolic consumption, the conventional thing becomes a symbol, and the equation—thing = symbol = meaning—is established. That is, the symbolic meaning and not the goods itself is emphasized. This can also be broadly interpreted as the lack of transparency of representation.

The modernity of the enlightenment loses reality owing to the lack of the transparency of representation and arrival of consumer society, and begins to embrace chaos. Postmodernity admits and embraces chaos. Michel Foucault advised that the difference against the uniformity and the fluidity against the system should be selected. The significance of postmodernity lies in allowing chaos and welcoming the "Tower of Babel"—it overthrows and eradicates the arrogance of modernity.

1.3.4 Dedifferentiation of the Functionalist Society

While the Modern Age seeks to advance society by making use of the concept of function primacy that emphasizes efficiency and rationality, its feature lies in accomplishing the highly functional differentiation. Max Weber called this the "rationalization process of society." On the contrary, postmodernism claims that dedifferentiation occurs by the very fact of the completion of the modern differentiation (Lash, 1990). For example, while authoritarian dual opposition such as man vs. woman, professional vs. layman, elite vs. the masses which modern society has created from the functional standpoint, is going to collapse at present; dedifferentiation shows that the movement needs to be considered from the beginning by melting such functional differentiation and removing their boundaries. Dedifferentiation does not imply a return to the undifferentiated state in the pre-Modern Age but the dedifferentiation that is caused by a result of highly successful differentiation.

In many cases, the modernist dichotomy tends to privilege one element over another. For example, the professional vs. layman argument inherently states that the layman should follow the suggestions of the professional, and the gender division of labor includes the ideology that women should do housework at home while men work outside. Dismantling this ladder is dedifferentiation; further, this seems to be a sociological version of deconstruction in poststructuralism.

In a consumer society, the commercialization of culture advances with the development of the culture industry and the distinction between high-grade culture and mass culture is unclear owing to the collapsing of the boundary between economy and culture. As an extension of the culture industry, art and music are advertised on television and stripped off their context and fragmented. Thus, the unity of their cultural meaning is destroyed. Although this phenomenon can be a criticized as the corruption of culture, such sorrow by itself represents a clinging to authoritarian differentiation in the Modern Age.

The discriminatory gender division of roles between man and woman also fluctuates according to the basis of the arguments made by feminists. Feminism has long sought a society wherein men and women are equal by campaigning for emancipation against gender discrimination. The fixed gender role differentiation that the man works outside the house and the woman manages household affairs has gradually dissolved. Further, parallel to this, androgynous (bisexuality) tendencies are emerging with the increasing feminization of men and masculinization of women. The careful application of makeup by men is also an example of dedifferentiation.

Although politics have been considered as activities belonging to the nation state, in the modern sense, many of the new social movements such as nuclear power plant contrarieties, movements for the protection of life and ecosystems, movements opposing environmental pollution, etc. are rarely guided by labor unions or political parties. New social movements, pertaining to crises in people's ways of life, are constantly seen at the civil society level. The idea that politics belongs to the nation state will thus collapse and be replaced by one which holds that politics belongs to civil society and daily life will increases.

In addition, as the borderless economy advances through the development of information networks, we cannot consider an economy and living standards only with respect to one country. The idea of considering the same within the framework of a nation will collapse rapidly owing to the borderless information network. Further, managing a country, as defined by the concept of sovereign nations since the nineteenth century, will gradually become impossible. Such borderless phenomena are also examples of dedifferentiation.

Moreover, the distinction between normality and abnormality will become complicated. Symptoms that cannot be judged as diseases or not increases, then these will be named syndromes. The distinction between nature and artificiality, too, will be complicated. Similarly, the distinction between day and night will disappear from the time zone of human life action, and the world will become a nightless city.

In short, the various distinctions that modern society has presumed will dissolve and dedifferentiation will occur. The very cause of such a situation is the differencification caused by rhizome. The conditions that allow the occurrence of dedifferentiation by rhizome are features of postmodernity. This phenomenon also occurs at the edge of chaos of the modern civilization. Therefore, questions pertaining to postmodernity offer the ideal opportunity for decoding contemporary age through the instability and the unpredictability of modernity itself to be an index, and giving up the promised future. Postmodernism is the trial that intends to get close to the critical point of modern civilization through the chaos and deconstruct it another civilization.

1.4 Self-Organization from the Edge of Chaos

The science of complex systems and postmodern theory are paradigms that promote the change from the order theory of the world to the chaos theory; these paradigms contribute to the perspective that foresees the self-organization of an emergent civilization by reexamining the modern civilization from the edge of chaos (Imada, 2001a, 2008, Chap. 4).

At present, the new current of the chaos theory of the world has arisen to the fore. This theory does not emphasize the conventional chaos that is opposed to order but highlights the edge of chaos located at the border between order and chaos. At the edge of the chaos of modern civilization, social systems go in an excited state, and the differencification dynamics of rhizome continually try to dedifferentiate the established parameters. Herein, cracks and distortions are generated in functional social structure, and noise and fluctuation occur frequently. Although noise and fluctuation correspond to the mutation in the biological evolution, these indicate the emergence of new information in social evolution.

Fluctuation and noise arising from dedifferentiation by rhizome are not the factors that lead the system to a crisis; rather, they drive the system into a different existence and structure. Although order theory considers such work to be accidental and random, the fluctuation located at the edge of chaos brings about a new order by amplifying itself. Here, individuals as micro units play a more important role than the macro units such as organizations and society. The new differences are articulated by the synergy of the micro units and order is established. Roles and statuses allocated with the functional requisite of the social system will be dedifferentiated, and the action that is not bound by them becomes dominant.

Moreover, the control becomes invalid at the edge of chaos of modernity. Therein, the existence of a control center and the analogue such as authorities and wholeness of the system are derived. This is the challenge that reverses the subordination of the individual for the whole, and leads to the trial that attempts to generate macroscopic order through the synergy of individuals without control center.

The stress on chaos and fluctuation apparently includes the danger of anti-scientism and mysticism that dissolves the world, life, and thought into randomness. However, the fluctuation at the edge of chaos should not be considered an accidental factor but an action of differencification and self-reflexion by the components (individuals) in order to find the new order under given conditions.

The increase of the arguments on complex systems and postmodernity symbolically indicates that modern civilization has reached the edge of chaos. A sketch of the new civilization will probably emerge from the chaos theory of the world. This implies the self-organization of the civilization scale, and from now on, the society will be in an excited state owing to the generation of noise, fluctuation, and chaos.

References

Ashby, W. R. (1956). *An introduction to cybernetics*. Chapman & Hall.
Axelrod, R. (1984). *The evolution of cooperation*. Basic Books.
von Bertalanffy, L. (1968). *General system theory: Foundations, development, applications*. George Braziller.
Deleuze, G., & Guattari, F. (1980). *Mille plateaux: Capitalisme et schizophrénie*. Minuit.
Derrida, J. (1972). *Positions*. Minuit.
Gleick, J. (1987). *Chaos: Making a new science*. Viking Penguin.
Haken, H. (1976). *Synergetics: An introduction, nonequilibrium phase transitions and self-organization in physics, chemistry and biology*. Springer Verlag. (1978, 2nd ed.).
von Hayek, F. A. (1973). *Law, legislation and liberty, vol. 1: Rules and order*. Chicago University Press.
Imada, T. (1986). *Jikosohikisei: Shakai Riron no Fukkatsu* (Self-organization: Revitalization of social theory). Sobun-sha.
Imada, T. (1987). *Modan no Datsu-kochiku: Sangyo Shakai no Yukue* (Deconstruction of the modern: Future of industrial society). Chuo Koron-sha.
Imada, T. (1994). *Konton no Chikara* (The power of chaos). Kodan-sha.
Imada, T. (1997). "Kanri kara Shien-e: Shakai Sisutemu no Kozo Tenkan wo Mezashite" (From control to support: Toward structural transformation of social system). *Soshiki Kagaku (Organizational Science), 30*(3), 4–15.
Imada, T. (2001a). *Imi no Bunmei-gaku Josetsu: Sono Saki no Kindai* (A discourse on the civilization of meaning: Modernity and beyond). University of Tokyo Press.
Imada, T. (2001b). "Fukuzatu Kei to Posutomodan: Jikososikisei Ron no Shiten Kara" (Complex systems and postmodern: From viewpoint of self-organization). In T. Imada et al. (Eds.), *Fukuzatu Kei wo Kangaeru: Jikososikisei to wa Nanika II* (Considering complex systems: What is self-organization II) (pp.6–45). Minerva Shobo.
Imada, T. (2008). *Self-organization and society*. Springer Verlag.
Imada, T. (2016). "Jikososhikisei to Shakai no Metamorufoze" (Self-organization and metamorphosis of society). In K. Endo et al (Eds.), *Shakai Riron no Saiko: Shakai Sisutemu-Ron to Saikiteki Jikososhiki-sei wo Koete* (The revival of social theory: Beyond social system theory and reflexive self-organization) (pp. 185–218). Minerva Shobo.
Jantsch, E. (1980). *The self-organizing universe: Scientific and human implications of the emerging paradigm of evolution*. Pergamon Press.
Kaneko, K., & Ikegami, T. (1998). *Fukuzatsu Kei no Shinka teki Shinario: Seimei no Hatten Yoshiki* (Evolutionary scenario of the complex systems: The development mode of the life). Asakura Shoten.
Kuroishi, S. (1991). *Sisutemu Shakaigaku: Ookisa no chi* (System sociology: Knowledge of size). Harvest-sha.
Langton, C. G. (1990). Computation at the edge of chaos: Phase transitions and emergent computation. *Physica D, 42*(1–3), 12–37.
Lash, S. (1990). *Sociology of postmodernism*. Routledge.
Levi-Strauss, C. (1958). *Anthropologie structurale*. Libraire Plon.
Lorenz, E. N. (1963). Deterministic nonperiodic flow. *Journal of the Atmospheric Sciences, 20*(2), 130–141.
Luhmann, N. (1972). *Rechtssoziologie, 2Bde*. Rowohlt Taschenbuch Verlag.
Luhmann, N. (1984). *Soziale Systeme: Grundriß einer allgemeinen Theorie*. Suhrkamp.
Lyotard, J.-F. (1979). *La condition postmoderne*. Minuit.
Nicolis, G., & Prigogine, I. (1977). *Self-organization in nonequilibrium systems: From dissipative structures to order through fluctuations*. John Wiley & Sons.

Packard, N. H. (1988). *Adaptation toward the edge of chaos*. University of Illinois at Urbana-Champaign, Center for Complex Systems Research.
Poster, M. (1990). *The mode of information: Poststructuralism and social context*. Polity Press.
Waldrop, M. (1992). *Complexity: The emerging science at the edge of order and chaos*. Simon & Schuster.

Chapter 2
A Retrospective on the University–Industry Innovation Nexus in Japan: Empirical Assessment of Coauthorship in the Light of New Data

Kenneth Pechter

Abstract The role of public research in contributing to the innovation that drives economic development is rightfully a topic of concern for policymakers throughout the world. The huge amount of open research taking place primarily in universities and public research institutes represents a major deployment of research investment, and raises the question of whether the research goes on to support innovation and economic development for the benefit of the people who support that research, and global society as a whole. This has particularly been an issue in Japan, where large public research investments coincided with the onset of prolonged economic recession, and where there has been a dearth of the lauded signs of the university–industry innovation nexus, such as university-originated spin-off companies or lucrative university-held patents. Supported by the lack of these signs, an assumption of a *disconnect* between university research and industrial development emerged as the conventional wisdom, upon which most research policy prescriptions have been based. With this as background, the author conducted a series of research studies from the mid 1990s to the early 2000s during the first of Japan's so-called "lost decades," and demonstrated empirically that the disconnect hypothesis was incorrect. In fact, the university–industry innovation nexus was quite strong as measured via industry support of university research, university–industry coauthored research, and economic development based on university-sourced patents. This finding—particularly the strong and growing trend of university–industry coauthorship of scientific publications over the 1981–1996 time period—was so contrarian to the conventional wisdom that skeptics doubted the veracity of the finding and questioned whether the trend was real or an aberration. Those doubts were debunked at the time based on the available data, but temporal distance would be required for a verification of the robustness in the time trend. With the benefit of two decades of hindsight and a secondary source dataset for the 1981–2004 time period, this new study was finally able to verify

K. Pechter (✉)
Hosei University Business School of Innovation Management, Tokyo, Japan
e-mail: pechter@ieee.org

K. Kijima et al. (eds.), *Systems Research II*, Translational Systems Sciences 27,
https://doi.org/10.1007/978-981-16-9941-2_2

the university–industry coauthorship metrics. The findings of this new study fully reconfirm the 1981–1996 results, and demonstrate empirically that the trend of rising university–industry coauthorship continued unabated through the 2004 end point of the new dataset.

Keywords National innovation systems · Economic development · Research policy · University–Industry linkage · Japan bubble economy

2.1 Introduction

In pop culture, politics, economics, and science & engineering, there was much ado about Japan's national innovation system during the meteoric rise and resounding pop of the Japanese economy in the final decades of the twentieth century. It was this context that first brought me to Japan in 1988 as a Ministry of Education scholar in the Takahara Lab of the Tokyo Institute of Technology, kept me in Japan as a junior faculty member in the Kodama Lab of the University of Tokyo's Research Center for Science and Technology, and carried me on further as a University of Tokyo professor, until I left the academia to pursue a career in industry in the mid-2000s.

The major result of my research back then was empirical measurement of a significant innovation nexus between the university and industry sectors in Japan. This was notable because the widely held conventional wisdom was that the Japanese innovation system suffered from a *disconnect* between academic research and industrial development due to the university sector's irrelevance to industry for anything other than as a supply of new employees.

Now back in academia as professor and Director of the Global MBA Program at Hosei University's Business School of Innovation Management, I look back on the key finding of that research—the empirical measure of university–industry coauthorship in scientific publications, which indicated a strong and rising linkage between the two sectors—and assess how the finding has stood up in the intervening years. In short, with the advantage of hindsight I attempt to answer the question that our contrarian result produced at the time, and remained until now: **Was the trend we found true, and if true would it hold up in the years to follow?**

I also take this opportunity to acknowledge the invaluable support I received from key people in the early part of my research career.

2.2 Why the University–Industry Innovation Nexus?

2.2.1 Global Context

After Japan's bubble economy deflated (at the end of the 1980s or further into the 1990s depending on what indicators are used to make the call), Japan entered an economic stagnation known as the *lost decade* that then turned into the lost decades, and continues into its third decade today according to certain measures (Morgan, 2021). Prior to the lost decades, consider how Japan's economy was both admired and feared. As a natural resource-poor nation with a small landmass, Japan rose out of its devastation in the Second World War via a non-orthodox capitalist approach—from the dominant US point-of-view—that included government intervention, industry coordination, and managed trade. Critics in the US and elsewhere leveled claims of corporate collusion, dumping of exports at below market costs, and free-riding on the basic research of other nations as reasons for Japan's success.

Against this backdrop, the university–industry nexus in innovation came into the spotlight. In the US, the 1990s saw economic output driven at least in part by investments in the university sector, including such famed cases as SUN Computers (**S**tanford **U**niversity **N**etwork), Google (founded by Stanford students Sergey Brin and Larry Page), the genetic revolution kicked off by the now COVID-famous PCR (polymerase chain reaction) at a University of California-related biotech startup, and a host of startup companies coming out of university-anchored regions like Silicon Valley, Route 128 in Boston, and the Research Triangle in North Carolina. The lack of similar university-originated economic success stories in Japan appeared to support the free-riding claim: **Japanese universities simply did not have good enough research to be of interest to industry other than as a supplier of new employees, and Japanese industry was able to make up for this deficiency due to its asymmetric access to the seeds of basic research in the US and elsewhere.**

And note, while such claims generally originated outside of Japan, observers in Japan did not raise strong disagreement with the university part of these claims, and even made innovation-related policy decisions with these claims as basis.

2.2.2 Personal Context

One effect of the Japanese economic miracle was that Japan became a subject of interest all over the globe. Popular and business press, arts including motion pictures and other mass entertainment, trade negotiations, and the attention of people in a variety of locations, vocations and avocations focused on Japan and its role in the world—this author included. As an Electrical and Computer Engineering undergrad in the mid-1980s at the University of California Irvine (at the time the American university with the second highest percentage of Asian students, following the University of Hawaii), and continuing with my advancement to UCI's Graduate School of Engineering to begin my Master's and PhD degree programs, the issue of Japanese competition in technology-based industries such as consumer electronics and autos hit close to home. But I was more bullish on American than fearful of Japan. Instead of thinking of the Japanese as the enemy and disparaging young people in America who no longer excelled in math and science (the term STEM was not yet born), it seemed to me that America's strength was not in topping international math exams but rather in adroitly finding new markets for new technology. Perhaps the US could learn something from rather than just vilify Japan.

Through good fortune, I had two UCI engineering professors with strong Japan connections. Professor Hideya Gamo was a Todai (University of Tokyo) physicist who was invited to be a researcher at IBM in the 1960s (along with future Noble laureate Reona Esaki and future lauded University of Illinois professor of VLSI system design Saburo Muroga). Professor Roland Schinzinger was born in Japan where his father Robert was a poet, philosopher, and professor of German literature. Incidentally, Roland earned the first UCI doctorate in engineering, under his graduate advisor and founding Dean of the UCI School of Engineering Robert Saunders—who I later found out was awarded a dissertation doctorate from the Tokyo Institute of Technology (more on this later).

Via Professors Gamo and Schinzinger and initially unbeknownst to me, I was nominated then selected as the first University California student out of the nine-campus UC system (now ten with the addition of UC Merced) to participate in a new graduate research exchange program with the Tokyo Institute of Technology. This was a few months into my Master's studies, and I quickly accepted what we expected to be a 6-month stay at TIT ("Japan's MIT," I was told). Before long, however, we found out that I would be receiving a Monbusho Scholarship (Japan's Ministry of Education Scholarship at the time, now called Monbukagakusho or Ministry of Education, Culture, Sports, Science, and Technology), which would require me to stay for 18 months. Since it would not make sense to have an 18-month break in the middle of a 20-month Master's program, I rushed to finish my UCI Electrical & Computer Engineering Master's degree before leaving for Tokyo in 1988 to work under the professor who generously agreed to accept me, Professor Yasuhiko Takahara in the Systems Science Department of TIT's Graduate School of Engineering.

Upon arrival in Japan, I will never forgot being picked up at the Yokohama City Air Terminal by a young TIT professor named Junichi Iijima, who brought me to my dormitory and kicked off the most amazing experience of my life. My 18-month stay at TIT was followed by 3 years of PhD study back at UCI while deep-diving into all things Japan (including a great deal of time under the tutelage of Professors Edward Fowler, James Fujii and Akemi Morioka in UCI's highly ranked East Asian Languages and Literature Department, to whom I am eternally grateful); a 1-year Japan Foundation Scholarship to study at the Inter-University Center in Yokohama under the leadership of Director Bruce Batten (I also receive tuition support due to recommendation by Tetsuo Nishide of Japan's Ministry of Economy, Trade, and Industry, the selfless Japanese bureaucrat who incidentally also worked to personally establish the California Institute of Technology Japan Internship Program that has brought Caltech students to work in Japanese companies since the mid-1990s); and withdrawal from my UCI PhD program and my Chancellor's Fellowship to accept a very junior faculty position at the University of Tokyo (a very difficult decision, made less scary due to the guidance of UCI Computer Science professor and founding Dean of the University of Michigan's School of Information, John Leslie King). This turned into 10 years on the Todai faculty, first as a research associate in the Kodama Lab at the Research Center for Advanced Science & Technology (while a faculty position, more like a research assistant in the US system), and then as associate professor and later professor at the Institute for Social Science and Graduate School of Interdisciplinary Information Studies. I then left academia to pursue what had brought me to Japan in the first place: To experience and learn from the practice of innovation in Japanese industry. It was during the 10 years I spent from 1995 to 2004 at the University of Tokyo that I did my work on the University–Industry Innovation Nexus in Japan.

2.3 University–Industry Coauthorship Linkage in Japan, 1981–1996

With the previous section as background, let us review my research findings up to where I left off 20 years ago at the opening of the twenty-first century.

2.3.1 Research Context

My research at that time came out of several of US–Japan related projects, in which Japan's national innovation system was central to the discussion. The first was as a research staff member of the Japanese side of a series of US–Japan Joint Task Forces, operated under the auspices of the US National Research Council (including the National Academy of Science, National Academy of Engineering, and the Institute of Medicine) and the Japan Society for the Promotion of Science's *Committee on Advanced Technology and the International Environment* (known by its chronological order among the university–industry cooperative research committees formed by the JSPS since 1933, *Committee 149).* I was given this opportunity by my direct superior on the University of Tokyo faculty (unlike the US, junior faculty in Japan do have direct superiors), Professor Fumio Kodama, who was the Principal Investigator of the Japanese team and one of the Directors of the overall project. My particular involvement focused on one of these task forces, the *Joint Task Force on Corporate Innovation.* One aspect of this experience that I still treasure was to attend meetings in Washington D.C. as a member of the Japanese side while the US side had as one of its leaders my former UCI professor F. Sherwood Rowland, a Nobel Chemistry Laureate who at the time was the foreign secretary of the US National Academy of Sciences.[1] The resulting report that summarized the findings

[1] Incidentally Professor Rowland had developed a close research relationship with Japanese researchers following his not well-known connection to the infamous Daigo Fukuryū Maru or Lucky Dragon 5 incident in which a Japanese tuna fishing boat with a crew of 23 men was contaminated by nuclear fallout from a United States nuclear weapon test at Bikini Atoll in 1954.

of the Task Force was published by the National Research Council, *New Strategies for New Challenges: Corporate Innovation in the United States and Japan* (United States-Japan Joint Task Force on Corporate Innovation, 1999).

The next project was a US–Japan research project titled *Universities and Science-Based Industrial Development* (USBID). This project was led by the overall direction of Professor Lewis Branscomb at Harvard University's Kennedy School of Government, Professor Richard Florida then of Carnegie Mellon University, and Professor Kodama, and was funded by the Japan Foundation's Center for Global Partnership. The resulting book that summarized the findings of this project was published by the MIT Press, *Industrializing Knowledge: University-Industry Linkages in Japan and the United States* (Branscomb et al., 1999).

The third project was my own research, leading up to and continuing beyond my University of Tokyo doctoral dissertation, "System Assessment for Innovation Policy Formation: Measuring the University-Industry Linkage in Japan and the United States" (Pechter, 2001c).

In the context of this essay, the key line of inquiry running through all three of these projects is the work I did on empirical measurement of university–industry research linkages with my collaborator at the time, Sumio Kakinuma of the Japan's National Center for Science Information Systems (NACSIS, now called NII, the National Institute of Informatics).

2.3.2 Conventional Wisdom of University–Industry Disconnect in Japan

The issue at stake was: In Japan's remarkable technology-fueled long-term economic growth in the postwar period, did the Japanese innovation system provide insights into an alternative approach to the US innovation system's prescriptions (which was in theory but less so in practice: competitive research and development free from industry collusion or government interference)? Or instead, did Japan's so-called economic miracle derive more from its second mover advantage, and its special Cold War relationship to the United States providing development benefits in the form of asymmetric assess to American research, transfer of US technology, and access to beneficial export markets? Honestly, the answer is some of both. But effective policy formulation demands more specificity than that.

To many key opinion leaders who held with the second half of the above question, Japan was a *free-rider* on American basic research as demonstrated by the perceived

---

It is also interesting to note that the career of UCI's other Nobel Laureate, my physics professor Frederick Reines—who was awarded for his co-detection of the neutrino and work on neutrino astronomy—was bookended by Japan connections: in the start through his work on the Manhattan Project developing the Hiroshima and Nagasaki bombs, and at the end by Japan's development of the Super-Kamioka Neutrino Detection Experiment for the purpose of observing supernovae.

lack of contribution to Japanese industrial development by Japan's own basic and public research. To these critics, Japan was able to accommodate its own basic research weakness by free-riding on research done in the US and elsewhere, and in the process not provide other nations similar access to its own research (the *asymmetric access* claim).

This view—widely held in the 1990s, and even taking root in popular culture though literature and film such as Michael Crichton's bestselling novel and hit film *Rising Sun*—was premised in both the linear model of innovation, and the partly apocryphal view that university research and university-originated startups drove much of the industrial development in the US innovation system. The linear model of innovation refers to a useful but not entirely sufficient view of the innovation process, in which the results of basic scientific research feed into technological development that in turn leads directly to new products and therefore economic returns.

There are many aspects of this model that match the empirical record. For instance, much was made at the time out of the revenue received by the University of California for its portfolio of university-derived patents. But the lion's share of this revenue came from a small number of hugely successful patents (for instance, polymerase chain reaction or PCR), with the rest of the portfolio arguably draining resources away from the performance of public science. (Is the time a professor spends dealing with the patent process worth the lost research time? Although the US sees huge innovation benefits from university-originated success cases, the answer is **No** more often than **Yes.**)

At the time, US industries from steel to consumer electronics to autos to computer chips were dominated by Japanese companies. This led to a renewed focus on regaining American competitiveness, in which the university–industry nexus was put forth as a key source of innovations via university-based patenting, university spin-off companies, and startup activity leveraging university research results.

The lack of similar signs of a vibrant university–industry nexus in Japan supported the view of Japan as a free-rider benefiting from asymmetric access to US basic research. There is no question that there has indeed been less university-derived startup activity in Japan as compared to the US. But addressing the reasons for this requires going beyond simplistic assumptions. The assumptions were simplistic in that critics did not account for the different status of university faculty in the US and Japan in regard to matters such as consulting, startups and patenting (keeping in mind that at the time, the strongest university research in Japan was done at national universities where at that time faculty were still part of the civil service and governed by civil servant regulations); simplistic data analysis that did not sufficiently account for measurement differences between the US and Japan in the collection and deployment of data such metrics as R&D expenditure; and insufficient treatment of the legal differences in the extent that academic involvement is disclosed on patents in the US and Japan.

One example of how frenzied the issue became is illustrated in reference to the aforementioned Crichton novel, *Rising Sun.* Crichton included UC Irvine in the novel as the setting where Hitachi had a corporate research lab into which only

Japanese passport holders were admitted entry—the point being that Hitachi was on campus to access university research, but the university community itself was not welcome into the Hitachi lab. The Hitachi Chemical Research Center was real. The Hitachi people explained to me that the Japanese passport thing was not true. It was just that as a corporation, like any corporation, you had to have business to be there. I had been invited in and do not have a Japanese passport. As I had the benefit of access to Mr. Crichton through my involvement in the Japan America Society of Southern California, I explained this to him when we met, and he told me that he would fix the misrepresentation in later editions. I never checked if he did.

The product of such a contentious environment was the widely held conventional wisdom of a disconnect between university research and industrial development in Japan. And perhaps the most surprising part of this story is that this even became the common view in Japan.

As a simple example, much Japanese innovation policy formulation in the 1990s was premised on the weak research links between Japanese universities and Japanese industry as demonstrated by the declining share of Japanese university R&D expenditure that was funded by Japanese industry (for example, Sakakibara & Ijichi, 2001). As good as the researchers doing the analyses were however—and they were indeed very good in general—they completely failed to consider that the decline was a result of the decline in total industry R&D spending following the collapse of the bubble economy, rather than a decline in interest in the university sector. We see this by looking from the perspective of industry at the share of total industry R&D spending going to the Japanese university sector, and observing quite easily that this percentage continued an upward trajectory signaling a rising not falling interest in the university sector (Pechter, 2001c).

In short, both inside and outside of Japan, the conventional wisdom was that Japanese university research was weak, and Japanese industry not interested in it.

2.3.3 Questioning the Conventional Wisdom of Disconnect

As mentioned earlier, my own experience in Japan started in 1988 in the laboratory of Professor Yasuhiko Takahara in the Systems Science Department of the Graduate School of Engineering at the Tokyo Institute of Technology. I was only there for about 18 months, but in that time I saw clearly that academic researchers were no strangers to industry. The Takahara Lab conducted sponsored research supported by industry funding; industry partners participated in research meetings; and the practice of long-term employment (known as *life-time employment,* in which a university graduate tends to retire from the company they entered when they graduated) meant industry researchers had a career-long span in which to nurture the relationship between their place of employment and their place of graduation. (I have not done the actual analysis, but my sense is that the majority of my TIT classmates back then still work in the company they joined when they graduated 30 years or so ago.)

Furthermore, as a student of Japanese language and culture starting in 1988, it did not take long for me to learn perhaps the first and most important rule for navigating Japanese society: human relationships are at the center of everything. For instance, even the simple act of presenting your business card involves an understanding of your company's level of prestige relative to the company of the person you are greeting, the relative hierarchical status of each of you in your respective companies, how much your own company has benefited or benefited from the other company, and an awareness of any other social connections that might exist between you and the other person. It did not seem to me that "disconnect" would be a likely state of the Japanese university–industry nexus.

Through another series of fortunate events involving the Takahara Lab—particularly the advice to seek out Professor Kodama that I received from Takahara-sensei's own sensei, Professor Takehiko Matsuda[2] —I ended up leaving my University of California Chancellor's Fellowship at the UCI, and becoming a faculty research associate at the Research Center for Advanced Science and Technology at the University of Tokyo, in the laboratory of Professor Kodama. Kodama, best known outside of Japan for his Harvard Business Review book *Emerging Patterns of Innovation: Sources of Japan's Technological Edge* (Kodama, 1995), studied innovation throughout his career from a particular perspective: empirical research. Kodama-sensei believed that no amount of talking about innovation could substitute for hard data on the subject. Under his guidance and throughout the three US–Japan innovation projects mentioned earlier, I was focused on finding ways to shed empirical light on the university–industry disconnect hypothesis in Japan. I eventually landed on three metrics of interest.

2.3.4 Measuring the University–Industry Linkage

For my dissertation research at the University of Tokyo—which I did while I was on the faculty—I focused on the three empirical metrics of R&D expenditure, patents, and scientific publication. All these metrics point to critical aspects of the innovation process. For each of these, it is possible to look from the perspective of the University–Industry Linkage (for instance, industry contribution to university R&D funding; university researcher contributions to patenting by industry; and university–industry coauthorship). And for each of these, given the conventional wisdom of a dysfunctional University–Industry Linkage in Japan, they had been insufficiently investigated (overly simplistic analyses of industry support of academic research and university involvement in industry patenting had already

[2] A student of Herbert Simon's at Carnegie Mellon University, a former president of TIT, a founder of the Operations Research Society of Japan, and, he later told me, that by coincidence he was closely involved in the awarding of the dissertation doctorate to UCI's founding Dean of Engineering, Professor Saunders, who I mentioned earlier.

concluded the linkage was broken, and insufficient attention had been paid to university–industry coauthorship).

Details on the results can be found in Pechter and Kakinuma (1999 MIT), Pechter and Kakinuma (1999), Kakinuma and Pechter (1999), Pechter (2000a, b, 2001a, b, c, 2002a, b, c). In short, however, for R&D expenditure and patents, our analyses indicated much stronger industry contribution to academic funding and more academic involvement in industry patenting than had been appreciated.

This essay specifically focuses on the third metric, coauthorship, since it was the finding of my earlier research that most challenged the conventional wisdom—and the one I have been wondering about for two decades since I left academic to pursue an industry career.

2.3.5 Main Findings of the 1981–1996 Dataset

The key finding of the earlier research with my collaborator Sumio Kakinuma was the substantial and increasing university–industry coauthorship ratio as seen from the industry perspective. We defined this ratio is the percentage of all research papers authored by someone in Japanese industry that also had coauthors from universities, whether the universities were in Japan or outside of Japan. In other words:

Japanese University-Industry Coauthorship Ratio
as seen from the Industry Perspective

$$= \frac{\text{Number of papers with a Japanese industry author coauthored with authors in universities of any country}}{\text{Number of papers with a Japanese industry author}}$$

The particular need for empirical analysis to inform the significant national innovation system-directed policy decisions being made at the time explains our focus on the *industry perspective.* Due to the conventional wisdom that Japanese university research was of minimal interest to Japanese industry, even those in Japan took this view as given with or without factual substantiation—and were introducing policy solutions to a possibly misdiagnosed problem. It was therefore imperative to generate empirical assessment of just how much university researchers were connected to the research and development activities of Japanese industry. **For this reason, in this essay, whenever the term university–industry coauthorship ratio is used, it should be understood that this refers to this *industry perspective*.** Our key finding of the university–industry coauthorship ratio is shown in Fig. 2.1.

For the reasons described above, this was a remarkable finding. Dr. Lewis Branscomb (former Director of the US National Institute of Standards and Technology, former Chief Scientist of IBM, former Chair of the Program on Science, Technology and Public Policy at the Belfer Center for Science and International Affairs in Harvard University's Kennedy School of Government), who was Co-

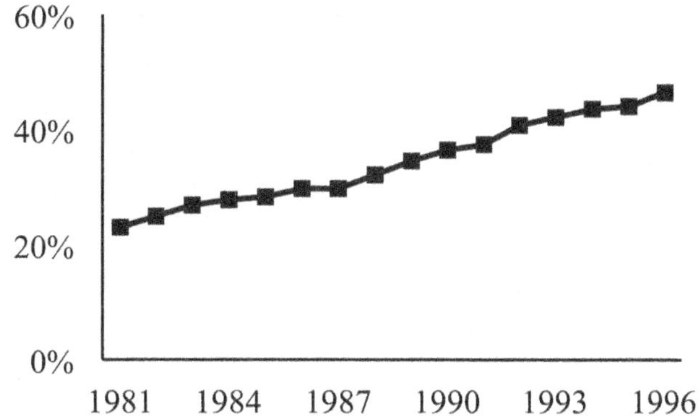

Fig. 2.1 University–industry coauthorship ratio in Japan, 1981–1996. (Source: original data, also in Pechter and Kakinuma (1999 MIT), Pechter and Kakinuma (1999), Pechter (2001c))

Principle Investigator on the Universities and Science-Based Industrial Development (USBID) Project that I co-directed, described the chapter written by my collaborator Sumio Kakinuma and me in our MIT Press book this way (Branscomb, 1999):

> This paper, entitled "Coauthorship Linkages between University Research and Japanese Industry," will have great impact on American thinking about Japanese innovation and the relationship of university research to industrial innovation in Japan. While American scholars point to the relatively low number of university patents licensed to industry and note that relatively little academic research in the national universities receives support from industry, they have missed the crucial evidence that Pechter and Kakinuma have uncovered. Their analysis of multi-authored Japanese technical literature finds that in papers with the first author from industry, the probability that one of the other authors is from a university is actually higher than in the United States. This observation and other data developed for this book will cause American politicians to rethink their position on the need for "symmetrical access" in US-Japanese technology relations.

For the reasons spelled out in our research papers, we chose *bibliometrics* (the application of statistical methods to the study of bibliographic data) as a prime empirical tool for illuminating the issue. Specifically, we analyzed the institutional affiliation at the level of university, industry, or other sector (primarily government research institutions) of the authors of scientific publications. We did this to characterize the university–industry linkage by analyzing events in which a paper published in the peer-reviewed science and engineering literature was submitted by at least one author affiliated to a Japanese industrial enterprise jointly with at least one author affiliated to a university (the university either inside or outside of Japan).

In order to assess the level of university–industry research interaction through bibliometric data, the analysis utilized the Science Citation Index database acquired from the Institute of Scientific Information (ISI) in the United States and maintained by the National Center for Science Information Systems (formerly NACSIS, now NII or National Institute of Informatics) in Japan. The database contained the set of Science Citation Index data for the years 1981–1996 in which at least one affiliation of a paper's authors is to an organization located in Japan. This set contained data

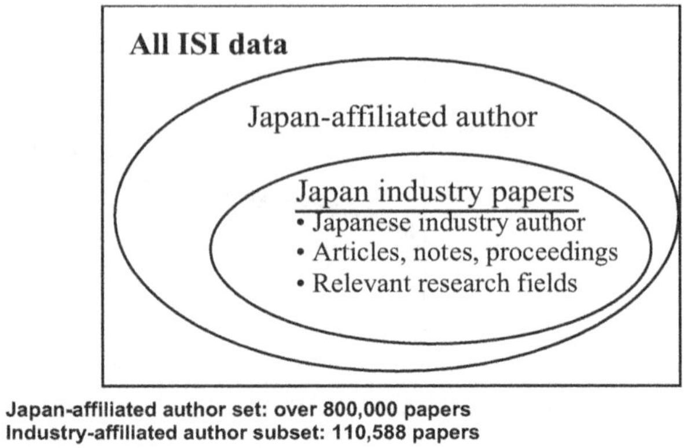

Fig. 2.2 Schematic of the bibliometric dataset in our earlier research. Japan-affiliated author set: over 800,000 papers. Industry-affiliated author subset: 110,588 papers

on approximately 800,000 papers. Note that this dataset came from the databases acquired and maintained by NII (NACSIS), and not from the web version.

Our analysis identified the industry subset out of these 800,000 papers according to three criteria:

1. Papers with at least one Japanese industry affiliation.
2. Papers categorized as *articles*, *notes*, or *proceedings papers* (the Science Citation Index also includes publications classified as *reviews*, *letters*, etc., but much of these are not the presentation of research).
3. Papers from the journals of all fields other than economics and business, education, law, psychology/psychiatry, and social sciences-general (the Science Citation Index classifies publications into fields of study on the basis of the specific journal in which each publication appeared).

This means that all papers without at least one Japanese industry author are immediately removed from the scope of analysis. The resulting study subset obtained based on these criteria contains 110,588 papers (Fig. 2.2).

With this data in hand, we were able to analyze key patterns in Japan's university–industry innovation nexus, revealed in the following figures.

The first figure shows the trend in number of papers. Note that based on the dataset, all of these papers have at least one author affiliated to an industry organization located in Japan, but the coauthoring organizations may be both domestic and international collaborators of the Japan-based firms (Fig. 2.3).

It is readily apparent from the figure the hat university–industry coauthorship mode of publication shows the strongest growth.

It is also useful to combine the inter-firm and intra-firm papers in order to make the data comparable to similarly disaggregated data on the US over the same time period, as shown in Figs. 2.4 and 2.5.

It is notable how similar the Japan and US data are, with the major difference being a slightly declining industry component in the US compared to a growing and then declining industry component in Japan (thought to be a result of the growth and then deflation of Japan's bubble economy).

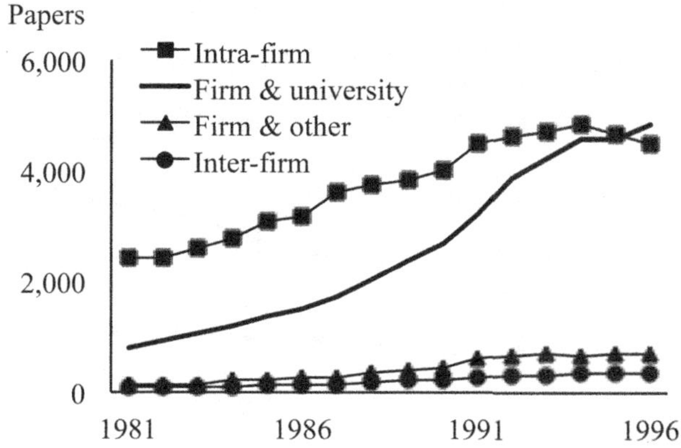

Fig. 2.3 Japan industry-authored papers, by mode of linkage. Intra-firm papers: by author(s) within a single firm. Inter-firm papers: by authors in different firms. Firm and university papers: by authors from both industry and university sectors. Firm and other: by authors in the industry sector and organizations in the "other" category (primarily national laboratories, public corporations and non-university hospitals). (Source: original data, also in Pechter and Kakinuma (1999 MIT), Pechter and Kakinuma (1999), Pechter (2001c))

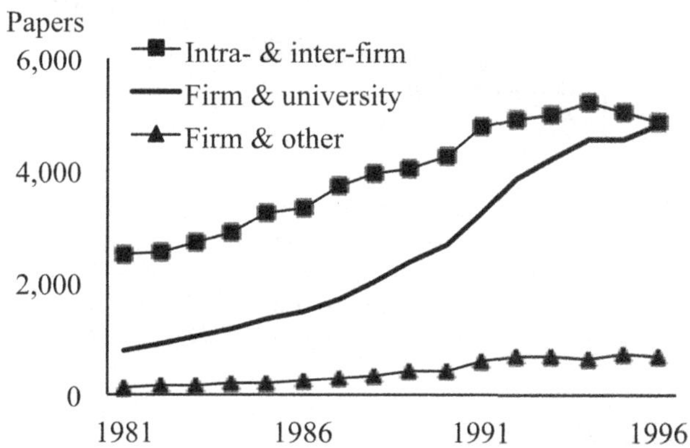

Fig. 2.4 Japan industry-authored papers, by mode of linkage (intra-firm and inter-firm combined). (Source: original data, also in Pechter and Kakinuma (1999 MIT), Pechter and Kakinuma (1999), Pechter (2001c))

Most surprising of all, however, was how closely our findings for the key university–industry coauthorship ratio for Japan matched other studies finding for the US (and also the UK for which we located comparable data). We see this in Fig. 2.6.

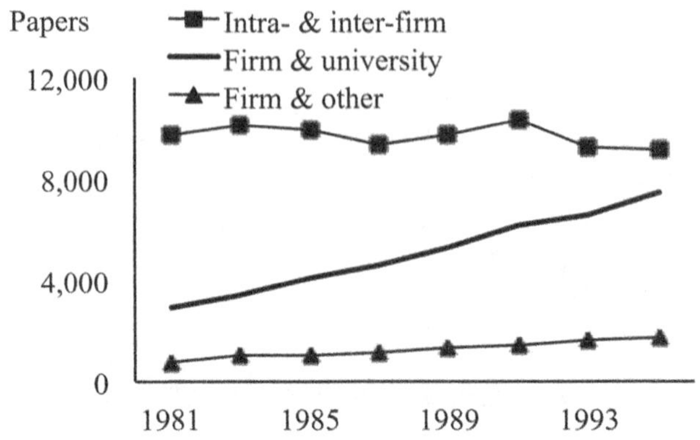

Fig. 2.5 US industry-authored papers, by mode of linkage (intra-firm and inter-firm combined). (Source: National Science Board (1998))

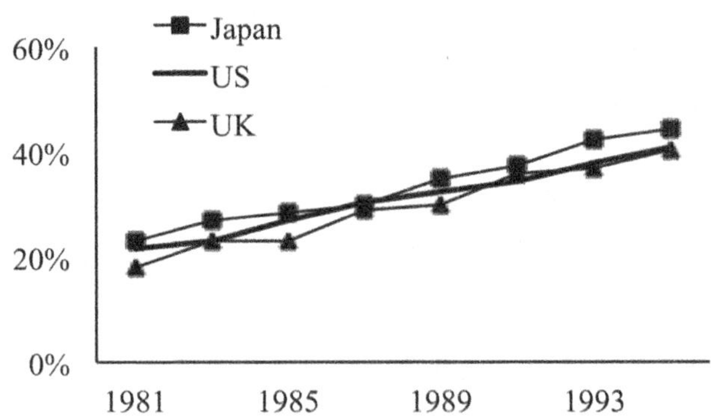

Fig. 2.6 University–industry coauthorship ratios in Japan, the US, and the UK. (Source: original data, also in Pechter (2002c), National Science Board (1998), Hicks and Katz (1997, p. 136))

2.3.6 *Skeptical Reaction*

Our finding went so much against the conventional wisdom at the time that it generated tremendous skepticism toward our conclusions. For instance, given the widespread acceptance of the prominent role of university research in the US innovation system, there was doubt that other countries—let alone Japan—would have similar coauthorship ratios. So, we found similar studies for the UK, and included that in our findings shown in Fig. 2.6.

We agreed that certain differences between the three studies of Japan, the US and the UK meant that the figures were not exactly comparable (the methodologies were very similar, but these were distinct and separate studies after all). In spite of that, the fact that all three started at comparable levels and grew dramatically over nearly two decades at essentially the same rate indicate that perhaps something more universal than the institutions of national innovation systems might be driving

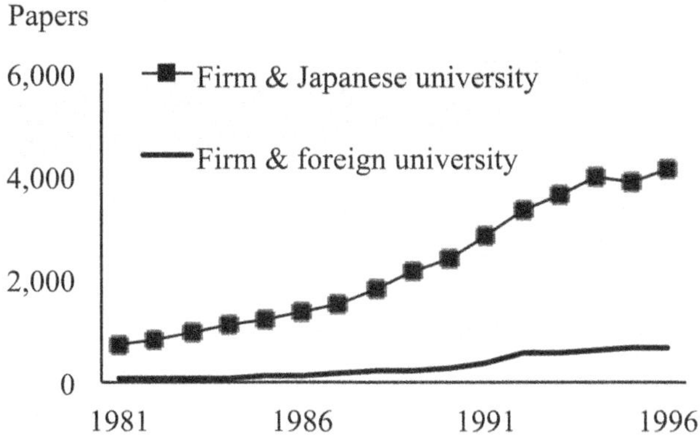

Fig. 2.7 University–industry coauthored papers in Japan shown, for Japan university coauthors only and foreign university coauthors only. (Source: original data, also in Pechter and Kakinuma (1999 MIT), Pechter and Kakinuma (1999), Pechter (2001c))

the change: namely, the patterns of innovation itself in the emerging paradigm of mega-competition and global business.

Another criticism was that the process of innovation may indeed be increasing the connection of industry to universities, but to American not Japanese university research (recall that while our dataset was limited to industry authors located in Japan, it included university authors located anywhere). So we performed the analysis shown in Fig. 2.7 as a rebuttal.

It is clear from the figure that university–industry coauthorship was not a result dominated by coauthorship with foreign universities. Consequently, even with the foreign component of Japanese university–industry coauthorship removed, the Japanese coauthorship ratio trend still closely matches that of the US (Pechter, 2000a).

The foreign university component was indeed growing, however. Understanding that this could be a very important trend in the changing nature of global innovation, we also looked at the foreign university component broken down by the G7 regions current at the time. The result is shown in Fig. 2.8.

As clearly seen in the figure, growth in coauthorship with foreign universities is not a story of growing dependence on American university research, which declined relative to other regions (and remained flat in terms of absolute number of coauthored papers in the 1990s).

Another claim leveled at the finding had to do with Japan's bubble economy. This claim said that the rise in university–industry coauthorship in Japan was a fleeting or even cosmetic affect of the bubble economy. It is important to understand that this explanation did not posit a change in the Japanese national innovation system in response to Japan's post-bubble economic transition as the driver of the change in coauthorship. Rather, the explanation posits that rising coauthorship was an unintended result of Japanese business practice under the influence of the

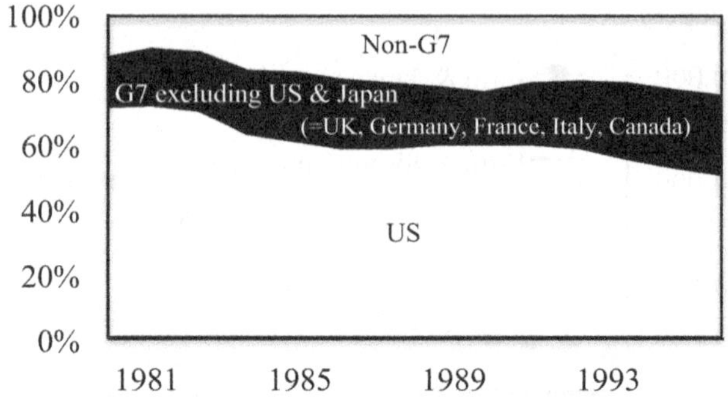

Fig. 2.8 Share of papers with an author in Japanese industry and a university coauthor located outside of Japan, by G7 country. (Source: original data, also in Pechter and Kakinuma (1999 MIT), Pechter and Kakinuma (1999), Pechter (2000a, 2001c))

superfluous economic environment during the bubble, much like the patterns seen in golf course membership turnovers and Japanese outbound tourism over the period.

This is a reasonable doubt to raise, but not a difficult one to answer. Earlier in this essay we discussed that the declining share of the Japanese industry-funded portion in total Japanese university R&D expenditure was erroneously taken as proof of a declining industry interest in university research in Japan, when this could be disproved simply by looking instead from the industry perspective (the increasing share of Japanese industry supported R&D expenditure that was transferred to Japanese university researchers out of total Japanese industry R&D expenditure). In the same spirit we applied a similar lens here and did not find any relationship between the growth of university–industry coauthorship in Japan and the bubble economy phenomenon. Details of this analysis can be found in Pechter (2000a).

The one caveat we made, however, was that at the time of the earlier research, we did not have enough temporal distance from the bubble economy to fully assess its effects. Time was required to gain the perspective to consider this possibility more fully.

The final claim made by critics as the "real" explanation for the rising university–industry coauthorship ratio found by our research was Japan's dissertation doctorate system. In Japan there are two procedures that may be used to obtain a doctorate. One is to enroll in and complete a PhD program in the graduate school of a university, which is generally speaking the same procedure commonly used in the United States. This procedure is known as the course doctorate, which refers to the fact that it is based on course instruction in a graduate program (and like the United States includes the writing of a dissertation). The second procedure is called the dissertation doctorate. Rather than enrolling in a graduate program, the candidate submits to a graduate school a dissertation based on research from their professional career for consideration as worthy of a doctorate. This latter procedure can be traced to the European education models Japan adopted in the late nineteenth century in which doctorates were awarded not for completion of a program but for professional research contributions to scholarly fields of study. This tradition can still be seen even in the United States in certain fields (e.g., some in the humanities) in which doctorates are awarded far into one's professional career. As such, the dissertation doctorate is not a shortcut to a PhD and has even been shown to be a higher standard than the course doctorate (Kodama & Nishigata, 1991).

Industry researchers, for whom world-class research facilities in the workplace and inflexible employment practices make the option of moving from firm to university less desirable, often use this procedure. While less so now in the twenty-first century, in the past many faculty in certain fields with a strong practical bent such as engineering never pursued doctorates before coming to academia, and these faculty often benefited from the dissertation doctorate system (such as UCI founding Dean of Engineering Saunders, mentioned earlier). The dissertation doctorate was also used more often in the past by faculty members of national universities, for whom civil service status prevents them from enrolling in course doctorate programs concurrently with being employed by their universities, which is why my University of Tokyo PhD was a dissertation doctorate (various changes in the legal status of both national universities and faculty have now reduced these restrictions).

Being unfamiliar, the dissertation doctorate system is seen as problematic and posited as the force behind the high and rising coauthorship ratio in Japan. Of course, compared to the high standards of doctorate systems in the US, Japanese doctorate systems may need to improve their quality control; but this is so for all doctorates, not just dissertation doctorates. If the claim were true, however, it could add a bias to the Japanese coauthorship ratio that would reduce the comparability between the coauthorship measures in Japan and the United States. Suffice it to say that our analysis did not support the dissertation doctorate system as an explanation for the rise in coauthorship ratio, as the number and trend of dissertation doctorates turned out to be uncorrelated with the rise in coauthorship (for details, see Pechter, 2000a).

In summary, the key finding of a large and rising university–industry coauthorship ratio in Japan, and the various additional empirical support for these findings, demonstrated clearly the falsity of the notion that Japanese industry has little interest in the research activities taking place in the Japanese university sector.

2.3.7 Unpursued Implications of Earlier Research

Our interpretation of these research findings at the time was that the widely held assumption of a disconnect in the innovation nexus between the university and industry sectors was incorrect. However, we did not conclude that this meant there was no problem with the Japanese university–industry nexus. At that time, Japan was well into its first *lost decade* of economic stagnation following the collapse of the economic bubble, and policy adjustments in the institutional relationships governing the nexus were needed. Any such adjustments, we argued, should be based on the empirical evidence, and not on assumptions.

In one of the studies that came out of this research (Pechter, 2002a), we posited that the problem in need of improvement was not the lack of university–industry linkages, but the lack of flexibility in these linkages. We cited the importance not only of efficient link formation, but also of the efficient severing of links as an imperative in the resource reallocation needed in support of innovation. In today's common parlance, we would refer to this as a need for disruption of the status quo as a prerequisite for innovation. At the time, however, the term "disruption" was not widely used (Clayton Christensen had just popularized it as an innovation term in his 1997 book *The Innovator's Dilemma* Christensen (1997)), and we ventured to use the unorthodox term "dis-organization" as a direct reference to the need to disrupt (to change the organizing forces in) the existing organizational structure. This key issue required further study.

We also noted that although the university–industry coauthorship ratio started out strong, grew strongly over the period of study, and was comparable to that in the US and UK—and was a robust finding in the face of all empirical scrutiny leveled against the finding—the data did indicate the possibility of a leveling off. If so, the leveling could herald of a coming decline in the coauthorship trend. This concern is seen most clearly in the last years of the time trend in the figure University–Industry Coauthorship Ratio in Japan, 1981–1996 (Fig. 2.1). Only time would tell, as additional years data became available.

This concern pointed to the future agenda for this research: **To study the trajectory of the university–industry nexus displayed in the university–industry coauthorship ratio in Japan, as it evolved in the latter half of the 1990s and moved into the twenty-first century.**

However, by that time I had been on the faculty of the University of Tokyo for about a decade, and I was driven by a desire to get back to the impetus for coming to Japan in the first place: To experience the practice of innovation in Japanese industry. By the early 2000s, the impact of Japan's burgeoning soft power was starting to emerge, a trend that would be dubbed "Japan's Gross National Cool" by Douglas McGray in 2002 (McGray, 2002), and later be adopted under the moniker "Cool Japan" by the Government of Japan. The global influence of Japan-originated media content including anime and video games, and the Japan-related content emanating from Hollywood and elsewhere in the form of titles such as *The Last Samurai, Memoirs of a Geisha, Kill Bill,* and *Lost in Translation* presaged what would gain traction as the Japan and Japan-relevant *content industries*. While these were fundamentally different from the science-based innovation of concern in my university–industry nexus research, I saw these emerging areas of economic activity as very much central to innovation. In fact, as the pursuit of profits based on the fruits of creative output—and using no small amount of technology in doing so—I saw these *creative* (or *cultural* or *Cool Japan*) industries at the time as very much innovation-based, and was drawn to the business sector to engage with them. So I left academia by the mid-2000s, and spent the next decade or so working with Japanese animation, film, television, entertainment, mobile, and internet media content, as well as fashion and advertising. For this reason, I never did follow up on the research agenda identified by my university–industry innovation nexus research:

specifically, the question of how the university–industry coauthorship ratio would evolve from there. **Would it continue to grow, or had we been witnessing the beginning of a decline in the linkage? Or were we simply incorrect to begin with, as many claimed?**

With this current essay, I am now able to consider this question, and provide an answer.

2.4 Post-1996 University–Industry Coauthorship Linkage in Japan

With the passage of time, I am now able to pick up where I left off two decades ago, and finally answer the question posed in the previous section: **Was our key finding of a strong and increasing university–industry coauthorship trend real—and did it persevere?**

A fuller account of the unfolding university–industry coauthorship linkage will require more in-depth primary source analysis, which means acquiring and processing a new dataset for the years since 1996. This is a huge and expensive undertaking. So for this initial survey, I will rely on some valuable key secondary source data, which we are fortunate to have thanks to the open model of scientific research publication.

2.4.1 Research Study of Interest, 1981–2004

The discussion of the university–industry innovation nexus within the functioning of national innovation systems has continued to be a subject of interest to researchers and policymakers in the two decades since I left off my own research, with three additional developments.

First, although the role of government (specifically, government-run research centers) has often been included in the analysis as it should be given its influence on innovation, the tools for managing such complex systems has advanced, making this much more effective. This was already happening in the 1990s when the Triple Helix Model of the knowledge-based economy (universities, industry, and government) was being developed by researchers such as Henry Etzkowitz and Loet Leydesdorff (for example, Etzkowitz & Leydesdorff, 1995, 2000). Since then, the approach has been broadened to utilize ideas from information theory to analyze mutual information in three dimensions as an indicator of Triple Helix relations at the systems level (Leydesdorff, 2003, 2006; Leydesdorff & Fritsch, 2006; Leydesdorff et al., 2006); and to consider more complex systems including additional sectors such as civil society institutions (Quadruple Helix), and the natural environment (Quintuple Helix) (Carayannis & Campbell, 2009, 2010; Leydesdorff, 2012).

Second, the globalization of innovation systems has taken a more central position in various studies, and the internationalization of both research and development has meant that more trans-national linkages have emerged, and in particular increased international collaboration via university–industry coauthored publications (for example, Igami et al., 2016).

Third, the relative position of nations has evolved. Japan's prolonged economic stagnation has reduced its position as China's tremendous economic growth has spurred on an active research portfolio. Such changes have been accompanied by variations in the position of nations, as major key research areas including AI and bio-medical technology have started producing results.

For the narrow purposes of this essay, however, we are interested in studies that have used comparable methodology to our original study, but extend beyond our earlier dataset's end date. **Namely, analysis of coauthorship data drawn from the bibliometric database of scientific publication going beyond the 1996 end date of our own study.**

Being that we are now in the decade of the 2020s, ideally we would like to find data sets that extend to the mid or even late 2010s if we were lucky enough to find such published data (keeping in mind that in the early 2000s the most recent accessible dataset went up to 1996). Currently, however, the best candidate dataset is from the Triple Helix work done by Leydesdorff and various colleagues, extending the coverage of the dataset up to 2004 (Leydesdorff & Sun, 2009). While this is short of our hoped-for ideal of extending through the mid-2010s, it does extend the window by 8 years. This is enough to see whether, as the tail end of our own research suggested, the trend of strong and growing connection of industry to Japanese academic research might be leveling off and perhaps even starting to decline. As this new dataset also starts in 1981, it provides a double-check of our own findings.

The research methodology of Leydesdorff and Sun is similar to that used in our earlier research. This is perhaps not so surprising, since there is widespread acceptance now of bibliometric analysis techniques based on citation index databases such as the former Institute of Scientific Information's (ISI) Science Citation Index and related databases for social sciences and arts & humanities etc.—now all under the umbrella of the Web of Science Core Collection maintained by Clarivate Analytics (former Intellectual Property and Science business of Thomson Reuters).

Moreover, Leydesdorff based much of his work on a Japanese dataset developed by a group of his collaborators in Japan, among them Negishi, Nishizawa, Watanabe and Sun (Sun et al. 2006, 2007). The Japanese research group worked to categorize all publications with at least one author address in Japan during the 1981–2004 period in the Science Citation Index, the Social Science Index, and the Arts & Humanities Index.[3]

[3] The methodological similarity to my own earlier research is not entirely coincidental, as my earlier work on coauthorship data would not have been possible with my close collaborator at the time, Sumio Kakinuma, who worked at National Center for Science Information Systems (NACSIS), the progenitor of National Institute of Informatics (NII). Professor Masamitsu Negishi

Table 2.1 Papers in Pechter & Kakinuma and Leydesdorff & Sun Datasets

Study	Pechter & Kakinuma (1981–1996)	Leydesdorff & Sun (1981–2004)
Number of papers—All Japan authors		
1981–1996 total	About 800,000	703,720
1981–2004 total	NA	1,277,030
Number of papers—Japan industry authors		
1981–1996 total	110,588	120,410
1981–2004 total	NA	205,634

Source: Pechter and Kakinuma (1999 MIT), Leydesdorff and Sun (2009)

The Japanese research group of Leydesdorff's dataset identified all publications in the three ISI databases that had at least one author address in Japan during the 1981–2004 period, and categorized authors into university, industry, government, or combinations of these three sectors, and also categorized the non-Japan authors addresses as foreign (but did not distinguish between types of foreign addresses, such as foreign university or foreign company, etc.). Note that by design of the dataset, foreign authors only were included in the dataset if they coauthored with an author in Japan.

This dataset consisted of 1,453,888 papers with at least one author address in Japan in the three ISI databases during 1981–2004. After performing standardization work to unify organizational name variations, and translation inconsistencies, organizational name changes and mergers and acquisitions etc. (Sun et al., 2007), the resulting final subset for the study contained 1,277,030 papers (87.8% of the original subset before standardization).

While this 1981–2004 dataset is quite similar in methodology to our earlier 1981–1996 dataset, there are differences of note:

- The two studies used different standardization processes to classify papers into university, industry, and other sectors.
- The 1981–2004 data contains only articles but uses all three ISI citation databases combined (the Science Citation Index, the Social Science Citation Index, and the Arts & Humanities Index), while the 1981–1996 used articles, notes, or proceedings papers, but excludes papers in journals classified as economics and business, education, law, psychology/psychiatry, and social sciences-general .

The resulting total number of papers in the two datasets is shown Table 2.1.

Because of the differences in the datasets, direct number-to-number comparisons must be made with these differences in mind. However, for such broad datasets occurring over a many years, component ratios (such as university–industry coauthorship ratio) and time trends do carry meaning.

---

also had some indirect involvement via his connection to Professor Kodama, to whose University of Tokyo laboratory I belonged.

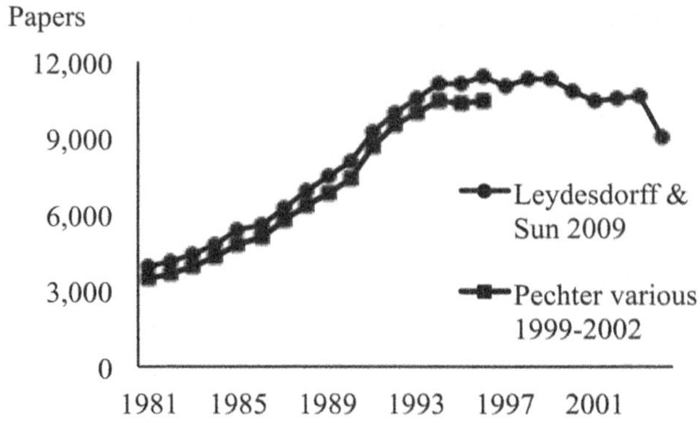

Fig. 2.9 Papers with an author from industry in Japan. (Source: original data, also in Pechter and Kakinuma (1999 MIT) and Pechter (2001c), reworking of data in Leydesdorff and Sun (2009))

2.4.2 Main Findings of the 1981–2004 Dataset

With the similarity and differences between the Leydesdorff dataset and our earlier dataset elucidated in the previous section, the first matter of interest is to compare the time series data for size (number of papers) of the industry papers between the two sets. We see that in Fig. 2.9.

As is clearly visible in the figure, the two datasets track each other quite closely. This is perhaps not surprising for the reasons of similarity explained in the previous section, but it is welcome confirmation of the earlier results. Our earlier dataset appears to undercount the industry papers relative to the newer dataset. Possible reasons for this include:

- The exclusion in the 1981–1996 dataset of non-science fields while the 1981–2004 dataset includes research from all fields.
- Techniques used to standardize records in terms of precise addresses for university and industry evolved (the latter study made further efforts to refine the process of accounting for institutional name variations and translation discrepancies when determining affiliations of authors) (Sun et al., 2007).

It is worth noting, however, that for the common years of 1981–1996, the industry paper count of our research was 110,588 and for the later research was 120,410, for a total coverage of 92% that improved over time.

For the university–industry coauthored papers we see an even closer result, as shown in Fig. 2.10.

For the common years of 1981–1996, the university–industry coauthored paper count (which includes papers that must have at least one Japan industry author, at least one university author either inside or outside of Japan, and could also have additional authors from non-industry and non-university sectors, primarily

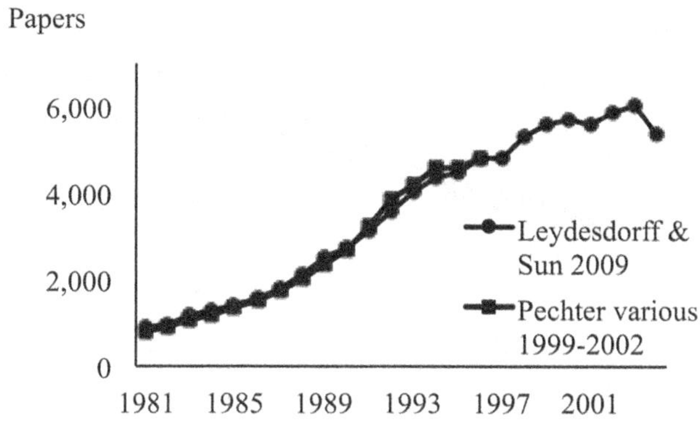

Fig. 2.10 Coauthored papers with an author from industry in Japan and a university author located anywhere. (Source: original data, also in Pechter and Kakinuma (1999 MIT) and Pechter (2001c), reworking of data in Leydesdorff and Sun (2009))

government research institutes) of our research was 40,903 and for the later research was 40,856, for a total coverage of virtually 100%.

With these welcome confirmations of our earlier research via the closeness of datasets, we are finally able to see how close our earlier results correspond to the latter results in terms of coauthorship ratio from the industry perspective.

It is important to note that looking from the perspective of industry was the intentional starting point of our earlier research. The reason is:

1. Given the conventional wisdom that Japanese university research of little interest to Japanese industry (widely accepted in Japan as well as overseas), it was first priority to examine the university–industry nexus from the point of view of industry.
2. The narrowing down of the required dataset and subsequent database management resources required were much more manageable for a small research project like ours if we started with the dataset of all Japanese industry-authored papers rather than all Japan-authored papers.
3. Given the nature of the science-oriented and/or public nature of university research, we do not expect industry support of research to be the major driving mode of research in the university sector. This supposition is borne out by the 1981–2004 dataset, for which the university–industry coauthorship ratio from the university perspective is calculable (in other words, the ratio of university–industry coauthored papers to all university papers in Japan), and indeed this ratio is generally less than 20% of the ratio from the industry perspective.

With this as preamble, comparison of the two datasets' industry perspectives on university–industry coauthorship (in other words, the ratio of university–industry coauthored papers to all industry papers in Japan) is shown in Fig. 2.11.

Inspecting first the 1981–1996 period common to both studies, the data agree well. While there is some slight divergence between our old data and the new data

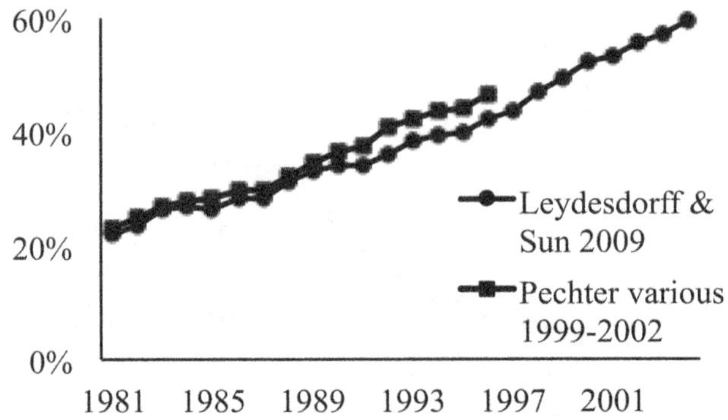

Fig. 2.11 University–industry coauthorship ratio for industry authors in Japan and university authors anywhere. (Source: original data, also in Pechter and Kakinuma (1999 MIT) and Pechter (2001c), reworking of data in Leydesdorff and Sun (2009))

(due to slight over counting of the old university–industry coauthorship data relative to the new data toward the end of the earlier study, as well as the under counting of the industry data relative to the new data toward the end of the earlier study as well), the time trends are remarkably similar.

This agreement is not surprising, as we already knew the both the ratio's numerator and denominator were similar between the two studies. The key point of interest is what happened next, in the period after the end of the 1981–1996 dataset study: **The university–industry coauthorship ratio continued to grow throughout the rest of the latter study period.**

In fact, the new trendline of the rise in the coauthorship ratio matches the earlier trendline, with both having a correlation of 99% to time (in other words, linear growth with time explains 98% of the change due to the *R*-squared, linear regression's coefficient of determination, equaling the square of the correlation in the case of two variables), and both the former trendline and the latter trendline has the same slope (increase rate of coauthorship ratio over time).

In the next two figures, we also see close agreement between the 1981–1996 dataset and the 1981–2004 dataset on the university–industry coauthorship ratio disaggregated into Japanese industry coauthorship with domestic Japanese universities and with foreign universities (Figs. 2.12 and 2.13).

Although we will discuss below that the increasing globalization in innovation systems may be leading to substantive changes in the patterns of domestic research interaction across the world, it is clear from the above two figures that trends of our earlier study have held steady beyond the 1996.

The sentence in the boldface type above is the number one result we have been wondering about for the past 20 years in response to our concern: **Did our finding of a significant and steadily increasing university–industry coauthorship ratio hold up; or did it change as a result of the dynamics of either Japan's national innovation system or emerging global patterns of innovation; or was our**

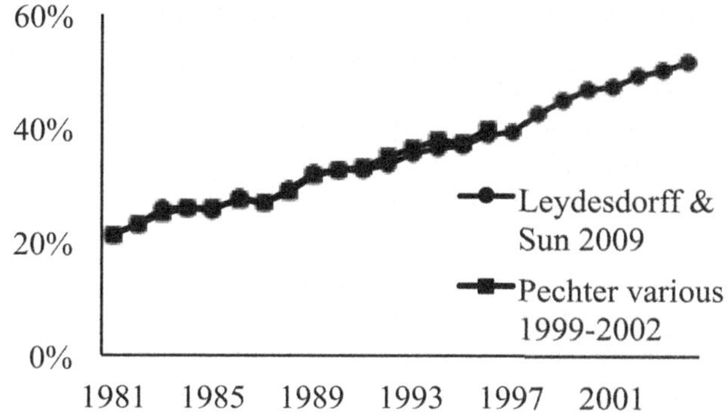

Fig. 2.12 University–industry coauthorship ratio for industry authors in Japan and university authors in Japan. (Source: original data, also in Pechter and Kakinuma (1999 MIT) and Pechter (2001c), reworking of data in Leydesdorff and Sun (2009))

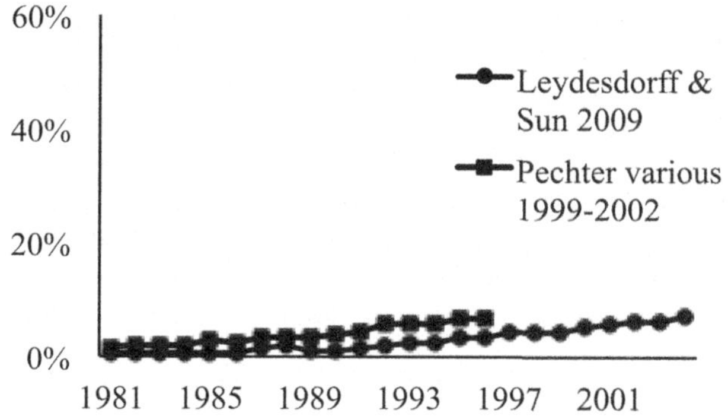

Fig. 2.13 University–industry coauthorship ratio for industry authors in Japan and university authors outside of Japan. (Source: original data, also in Pechter and Kakinuma (1999 MIT) and Pechter (2001c), reworking of data in Leydesdorff and Sun (2009))

finding even perhaps flawed somehow which would be revealed in a failure to be confirmed by later studies?

At last, we now have the answer (of course, since the secondary source dataset was from 10 years ago, we did not have to wait 20 years to find it).

2.4.3 Implications of the New Study on Our Earlier Study

With the welcomed confirmation of our earlier finding, I would like to delve deeper into the confirmation to the extent allowed by the newer dataset, and consider more of the findings of our earlier research. In particular, the newer study contains data that sheds light on some of the issues mentioned above.

We already covered the biggest concern: Was the slight weakening in the growth of the university–industry coauthorship ratio the beginning of the end of the steady upward time trend? The answer is a clear **No**.

The more recent study also sheds light on the matter of dependence on foreign university research. We had concluded 20 years ago that while the connection to foreign university research was growing, it was still small compared to the connection to domestic university research (less than 10% of the total connection). Of course, we expect more domestic research connections than cross-border connections, so the trend in a growing connection with foreign researchers is important. However, the newer data indicates that as of 10 years ago toward the end of this study, the foreign university connection of Japanese industry was still just over 10% of the total university connection.

It is important to keep in mind that one of the most important findings of the bibliometric research conducted in the last two decades is an increase in international coauthorship of research papers (note that coauthorship here refers to trans-national coauthorship by researchers in different countries, and does not necessarily refer to university–industry coauthorship) (for example, Igami et al., 2016).

This trend was also seen in the Leydesdorff and Sun (2009) research. Over the 1981–1996 period of our earlier study, the Leydesdorff and Sun study shows the percentage of all Japanese papers with an overseas coauthor going from 5% to 14% (note that since our original study started with the dataset of industry-authored papers only, a comparable figure does not exist). In the 8 years following the end of our earlier study, the 14% of 1996 continued upward to 22% in 2004. The globalization of national innovations systems is a key agenda item for further research.

2.5 Concluding and Looking Forward

This essay's simple purpose was to empirically assess the metric of university–industry coauthorship in Japan, extending beyond the 1996 endpoint of my earlier study of the same metric. The reason for doing this was because the key finding of my earlier research—a significant and rising university–industry coauthorship ratio in Japan—was so contrarian to the conventional wisdom of the time that it called into question the veracity of our analysis. Time was required for enough additional years' data to accumulate in order to confirm or contradict our finding—and for someone to do the requisite analysis, as I had transitioned from an academic career to a business career and did not do the analysis myself.

The conventional wisdom of the time and still persisting now, is that Japanese public sector and basic research is weak, and therefore Japanese industry is more interested in accessing the open research in other countries such as the US; as a result, Japan is free-riding on the basic research done elsewhere without providing similar access to valuable research in its own innovation system. There is no doubt that basic research in Japan may call for various improvements (for instance, in the past 10 years Japanese R&D spending growth has lagged (OECD, 2021), Japan has experienced a decline science publications (National Science Board, 2018, Chap. 5; OECD, 2017), and Japan's number of scientific publications with high citations has moved from ninth to tenth in the world, while China surpassed the United States for the first time and ranked first among major countries in scientific publications with high citations (NISTEP, 2021)).

Care must be exercised, however, when interpreting citation counts. For instance, one of the top takeaways from Stanford University's *Artificial Intelligence Index Report 2021* was that although China overtook the US in AI journal citations (after surpassing the United States in the total number of AI journal publications around 2016), "the United States has consistently (and significantly) more AI conference papers (which are also more heavily cited) than China over the last decade" (AI Index Steering Committee, 2021). Important nuances lie below the surface of macro trends.

Although the brash assertions of asymmetric access were based in part on faulty grounds, there may indeed be something to the issue of lack of access to Japan. The problem, however, may be more complex that a simple lack of things worth accessing. For instance, in a recent study, Japan was ranked 196 out of 196 countries in the ratio of inward Foreign Direct Investment to GDP, just behind North Korea. As journalist and highly regarded Japan analyst Richard Katz has pointed out, the cabinet-level council tasked with promoting Foreign Direct Investment (FDI) in Japan acts as if the problem is a lack of companies in Japan that could attract the interest of foreign investors, when in fact such companies are not few. As a result, the council's main policies are aimed at increasing attractiveness rather than at addressing the actual and more complex obstacles (Katz, 2021). The parallels between the problems of Japan's FDI policies and university–industry linkage policies (both which take lack of anything worth accessing as the obstacles to growth) are striking, down to even the imprudent use of metrics (the insufficient treatment of statistics for making international comparisons in both cases) (Ibid).

There are a variety of factors that may be inhibiting Japan's innovation system from more effectively engaging with the global economy and reviving its long-stagnant economy (long-time Japan watcher William Pesek for instance cites "cutting red tape, modernizing labor markets, catalyzing a startup boom, empowering women, boosting productivity and attracting international talent" (Pesek, 2021)). Addressing such issues requires looking beyond simplistic assumptions and grasping the state of the system more realistically. That was the imperative of my earlier research: To start at the most basic level and actually measure the university–industry linkage in research.

This earlier research clearly showed a strong, growing, and internationally comparable university–industry linkage in Japan. And subsequent analysis at the time rebutted conventional wisdom derived counter-explanations such as that the university–industry linkage we measured was a cosmetic result of the bubble economy, a result of dependence on American university research, or a side effect of Japan's dissertation doctorate system. Even so, we were eager for other research groups to validate or contradict our findings. This required us to wait and see. One more counter-explanation was that our finding was a temporary trend that was about to end. The only way to assess this claim as well was to wait for time to pass and see what happened in the years to come.

With this essay, we finally were able to examine the research record for other research that could check our results. We were pleasantly surprised that the record supported our findings tremendously well, with no detraction from our conclusions. The university–industry linkage as measured by coauthorship of scientific publication was as strong as we said, rose as much as we said, and this trend continued for the additional 8 years that the new study covered (Leydesdorff & Sun, 2009).

While 8 years is a big extension of the endpoint past 1996, 2004 was already 17 years ago. A good next step would be to find more recent secondary analysis of comparable datasets, or to invest the substantial time and money to generate the datasets and conduct primary analysis.

My earlier research also analyzed university–industry innovation nexus in terms of cross-sector R&D expenditure flows and cross-sector contribution to patenting. An extension to this analysis would also be valuable.

Moreover, my original research position was not that the university–industry nexus in the Japanese innovation system was fine and in need of no improvement. My intent was to remove spurious assumptions so that issues in need of attention could be addressed. The hypothesis at the time—which I still believe is relevant today—is that the problem with the Japanese university–industry linkage was not one of disconnect, but rather of not enough flexibility in the connections. The efficient reallocation of resources demanded by innovation requires long-term stable relationships be disrupted (at the time I used the term "dis-organized" (Pechter, 2002a, c)). More work is needed to assess this problem and generate policy recommendations.

For example, we can look at the Leydesdorff and Sun (2009) dataset to which this essay owes its existence. The Leydesdorff group using its own Triple Helix methodology, incorporating Shannon-type mutual information theory, had cautionary results for the Japanese university–industry innovation nexus. They found that increasing internationalization of the innovation system in Japan is changing the nature of linkages between the university, industry, and government research sectors in Japan. In particular, "a major trend is increased internationalization of academic research" resulting in "a relative decline of university–industry relations in Japan."

This said, according to the Stanford AI report, in terms of joint university–industry peer-reviewed AI Japan publications, Japan publishes far less in AI fields than either the US or China (adjusting for size of their economies, Japan and China are similar to each other, and are about 75% of the US). But in terms of the gold standard citation metric of Field-Weighted Citation Impact (Field-Weighted Citation Impact is the ratio of the total citations received divided by the total citations that would be expected based on the average of the subject field, with 1 designating meets expected citation level, and less than 1 or greater than 1 accordingly indicating below- or above-average performance), China and Japan are both about 1.3 while the US is about 2.1 (AI Index Steering Committee, 2021). It would appear that although Japan lags in absolute amount of university–industry research impact in AI, it is carrying its own weight adjusting for economic size.

More work needs to be done in improving our view of the nature of national and global innovation systems in general, and in assessing the Japanese system in order to support effective policymaking. But for the immediate purpose of this essay, verifying our earlier key result of a strong and increasing university–industry coauthorship linkage—and confirming that it held up over time—is a welcome and long-awaited confirmation. And recognizing some of the key people who made this work possible was also long overdue.

References

AI Index Steering Committee. (2021). *The AI index 2021 annual report*. Human-Centered AI Institute, Stanford University, Stanford.

Branscomb, L. (1999, January 8). Official correspondence to University of Tokyo.

Branscomb, L. M., Kodama, F., & Florida, R. (Eds.). (1999). *Industrializing knowledge: University-industry linkages in Japan and the United States*. MIT Press.

Carayannis, E. G., & Campbell, D. F. J. (2009). 'Mode 3' and 'Quadruple Helix': Toward a 21st century fractal innovation ecosystem. *International Journal of Technology Management, 46*(3), 201–234.

Carayannis, E. G., & Campbell, D. F. J. (2010). Triple Helix, Quadruple Helix and Quintuple Helix and how do knowledge, innovation, and environment relate to each other? *International Journal of Social Ecology and Sustainable Development, 1*(1), 41–69.

Christensen, C. M. (1997). *The innovator's dilemma: When new technologies cause great firms to fail*. Harvard Business School Press.

Etzkowitz, H., & Leydesdorff, L. (1995). The Triple Helix—University-industry-government relations: A laboratory for knowledge based economic development. *EASST Review, 14*, 14–19.

Etzkowitz, H., & Leydesdorff, L. (2000). The dynamics of innovation: From national systems and 'mode-2' to a TRIPLE Helix of university-industry-government relations. *Research Policy, 29*(2), 109–123.

Hicks, D., & Katz, S. (1997). A national research network viewed from an industrial perspective. *Revue D'Economie Industrielle, 79*(1), 129–142.

Igami, M., Saka, A., & Sadao, N. (2016). Situation of international collaborations in Japan—Evidence from bibliometric analyses and surveys to authors of scientific papers (in Japanese, Nihon no KokusaiKyoudouKenkyuu no Joukyou – RonbunBunseki to RonbunKyouchosha he no ShitsumonHyouChousa kara no Shisa). *The Journal of Science Policy and Research Management* (in Japanese, Kenkyuu Gijutsu Keikaku), *31*(2), 145–158.

Kakinuma, S, & Pechter, K. (1999). University-industry research collaboration in Japan: Industry dependence on academic research (in Japanese "Nihon Kigyou to Daigaku no Kyoudou Kenkyuu - Daigaku Kenkyuu e no Izon"). *Research Bulletin of the National Center for Science Information Systems* (in Japanese *Gakujutsu Joho Senta- Kiyo), 10.*

Katz, R. (2021, August 2). Japan as 196th. *The Oriental Economist.*

Kodama, F. (1995). *Emerging patterns of innovation: Sources of Japan's technological edge.* Harvard Business Review Press.

Kodama, F., & Nishigata, C. (1991). Structural changes in the Japanese supply/employment systems of engineers: Are we losing or gaining? In D. S. Zinberg (Ed.), *The changing university* (pp. 101–128). Kluwer Academic Publishers.

Leydesdorff, L. (2003). The mutual information of university-industry-government relations: An indicator of the triple helix dynamics. *Scientometrics, 58*(2), 445–467.

Leydesdorff, L. (2006). *The knowledge-based economy: Modeled, measured, simulated*. Universal Publishers.

Leydesdorff, L. (2012). The Triple Helix, Quadruple Helix, ..., and an N-tuple of helices: Explanatory models for analyzing the knowledge-based economy? *Journal of the Knowledge Economy, 3*(1), 25–35. https://doi.org/10.1007/s13132-13011-10049-13134

Leydesdorff, L., Dolfsma, W., & Van der Panne, G. (2006). Measuring the knowledge base of an economy in terms of triple-helix relations among "technology, organization, and territory". *Research Policy, 35*(2), 181–199.

Leydesdorff, L., & Fritsch, M. (2006). Measuring the knowledge base of regional innovation systems in Germany in terms of a Triple Helix dynamics. *Research Policy, 35*(10), 1538–1553.

Leydesdorff, L., & Sun, Y. (2009). National and international dimensions of the Triple Helix in Japan: University-industry-government versus international co-authorship relations. *Journal of the American Society for Information Science and Technology, 60*(4), 778–788.

McGray, D. (2002, May–June). Japan's gross national cool. *Foreign Policy* (pp. 44–54).

Morgan, J. (2021). The end of the Abe Administration—The end of abenomics? Books on past and present in the Japanese political economy. *Quarterly Journal of Austrian Economics, 24*(1), 167–187.

National Science Board. (1998). *Science & engineering indicators*. National Science Board.

National Science Board. (2018). *Science & engineering indicators*. National Science Board.

NISTEP. (2021). *Japanese science and technology indicators 2021*. Center for S&T Foresight and Indicators, National Institute of Science and Technology Policy (NISTEP), MEXT.

OECD. (2017). The digital transformation (science, technology and industry scoreboard 2017).

OECD. (2021, March). Main science and technology indicators.

Pechter, K. (2000a). Empirical support for significant university-industry linkage in Japan. In *Proceedings of the International Conference on the Management of Innovation and Technology*, IEEE Engineering Management Society, Singapore, November 13–14.

Pechter, K. (2000b). Understanding and communicating needs and solutions: A case of comparative policy analysis. *Social Science Japan, 18*, 27–31.

Pechter, K. (2001a). Japanese innovation reform in the light of past dialogue: Conceptions of convergence as perspectives for comparative system assessment. *Comparing Economic Institutions*. In *Proceedings of the 2001 Conference of the European Association for Evolutionary Political Economy*, Siena, Italy, November 8–11.

Pechter, K. (2001b). Comparative policy analysis under conditions of innovation-driven change: Assessment of the university-industry linkage in Japan and the United States. *Change Management and the New Industrial Revolution*. In *Proceedings of the 2001 IEEE Engineering Management Society International Engineering Management Conference*, Albany, New York, October 7–9.

Pechter, K. (2001c). *System assessment for innovation policy formation: Measuring the university-industry linkage in Japan and the United States* (Unpublished Doctoral Dissertation). The University of Tokyo.

Pechter, K. (2002a). Dis-organizing structure as a policy reform objective: Public-private networks in Japan's innovation system. *Managing Technology for the New Economy*. In *IEEE Engineering Management Society 2002 International Engineering Management Conference*, St. John's College, Cambridge, UK, August 18–20.

Pechter, K. (2002b). Field specialization and sectoral division of research in National Innovation Systems. In Q. Xu, X. Wu, & J. Chen (Eds.), *Management of technology and innovation in the 21st century* (Vol. 2002, pp. 62–67). Zhejiang University Press.

Pechter, K. (2002c). Comparative analysis of inter-sector innovation enterprises: The case of university-industry linkage reform in Japan and the United States, Joint Princeton-Northwestern Junior Scholar's workshop on embedded enterprise in comparative perspective, Princeton University, April 11–14, 2002.

Pechter, K., & Kakinuma, S. (1999). Coauthorship as a measure of university-industry research interaction in Japan. *Technology and Innovation Management*. In *Portland International Conference on Management of Engineering and Technology* (pp. 87–93).

Pechter, K., & Kakinuma, S. (1999 MIT). Coauthorship Linkages between university research and Japanese industry. In L. M. Branscomb et al. (Eds.), Industrializing knowledge: University-industry linkages in Japan and the United States (pp. 102–127). MIT Press.
Pesek, W. (2021, September 6). Japan must look beyond the usual suspects for its next leader—Suga's departure is a golden opportunity to reset the economy, Nikkei Asia.
Sakakibara, S. & Ijichi, T. (2001). The state of the university-industry linkage in Japan and the problem of conflicts of interest (in Japanese "Nihon ni okeru san-Gaku Renkei no Jittai to Rieki Souhan Mondai"). In M. Aoki et al. (Eds.), *University Reform: Issues and Discussion* (in Japanese Daigaku Kaikaku no Kadai to Souten, pp. 369–392). Toyo Keizai Inc.
Sun, Y., Negishi, M., Nishizawa, M., & Watanabe, K. (2006). Research linkage between university and industry in Japan: Comparison based on NCR-J and CJP databases. *Journal of Japan Society of Information and Knowledge, 16*(2), 7–12.
Sun, Y., Negishi, M., & Nishizawa, M. (2007). Co-authorship linkages between universities and industry in Japan. *Research Evaluation, 16*(4), 299–309.
United States-Japan Joint Task Force on Corporate Innovation. (1999). *New strategies for new challenges: Corporate innovation in the United States and Japan*. National Research Council/National Academy Press.

Chapter 3
Consideration of Organization Model Based on Dynamic Equilibrium Theory

Hirolkazu Tanaka

Abstract The external environment of firms is constantly changing. To adapt to change, they are required to continue to change their internal environment. In this chapter, the author considers the coordination mechanisms existing in constitutive elements of business organizations, applying the concept of Dynamic Equilibrium Theory in the field of life science.

Inflexible mechanisms will not affect a short-term business performance, whereas will endanger a firm's existence in the long term. To avoid such a risk, sophistication of coordination mechanisms should be highly required. This chapter also discusses measures to advance coordination mechanisms.

Keywords Dynamic capabilities · Coordination mechanisms · Dynamic equilibrium · Organizational change

3.1 Introduction

Firms constantly face minor and major changes in their external environments. A business cannot endure unless it can transform and adapt to changes.

The dynamic capability theory, which deals an organization's adaptability from a resource-based theory perspective, has been attracting recent attention. Its proponents, Teece et al. originally defined the concept as "the firm's ability to integrate, build, and reconfigure internal and external competences to address rapidly changing environments" (Teece et al., 1997). The dynamic capabilities approach was expanded to accept the idea that a firm's capabilities for change are restricted by path dependencies, rigidity, and its adoption of the organizational learning concept (Teece, 2007, 2012).

H. Tanaka (✉)
Shizuoka University, Shizuoka, Shizuoka, Japan
e-mail: tanakah@inf.shizuoka.ac.jp

K. Kijima et al. (eds.), *Systems Research II*, Translational Systems Sciences 27,
https://doi.org/10.1007/978-981-16-9941-2_3

In contrast, Zollo and Winter studied dynamic capabilities from an evolutionary economics perspective. They define dynamic capabilities as "a learned and stable pattern of collective activity through which an organization systematically generates and modifies its operating routines in pursuit of improved effectiveness (Zollo & Winter, 2002)"

There is also much ambiguity regarding the dynamic capabilities' locations, whether management will implement them or whether the actions of the site will spread throughout the organization (Huang, 2011).

As described above, the dynamic capability theory has been studied from various viewpoints, but its essence is not yet well understood.

This chapter introduces the dynamic equilibrium theory in life science based on Zollo and Winter's argument that the source of dynamic capabilities lies in organizational routines. It further considers an organizational transformation model based on this theory.

3.2 Concept of Dynamic Equilibrium

In a life science approach, Fukuoka states, "recombinant DNA techniques and approaches for elucidation of mechanisms of diseases view 'life' as a kind of machine with multiple components causally linked together (Fukuoka, 2011). Comparatively, dynamic equilibrium focuses on ceaseless movements of living organisms with a flow of decomposition and generation while internalizing external substances in order to maintain the order of life." Therefore, dynamic equilibrium is characterized by the relationships and complementarity between the components of life (Fukuoka, 2009).

Complementarity refers to keeping a system in balance by replacing one component with another when its function is lost or reduced. Thus, the components of life are always updating to create a new state of equilibrium, even in a changing environment (Fukuoka, 2011, 2017).

3.3 Organizational Model of Dynamic Equilibrium

From a micro viewpoint, a business's daily operations involve constantly processing customer orders (orders). A business organization's "flow of resources" in its environment mimics the dynamic equilibrium concept of living organisms—that a business organization is formed of multiple components and has coordination mechanisms to maintain a state of equilibrium among them.

The decomposition and generation of components are generated autonomously in the flow. In this chapter, the author examines the coordination mechanisms of three major components: people, the organization, and business processes. Each component consists of two layers: the surface layer and the deep layer. These components are described in the following sections.

3.3.1 *People*

People are committed to their organization by what Bernard called, "incentives and contributions." Their competence (knowledge, judgment, planning, negotiation and leadership) and attitude (discipline, accountability, aggressiveness and collaboration) must be latent in order to perform "work" that contributes to the organization. "Competence and attitude" are the key enablers for "work" (See Fig. 3.1).

3.3.2 *Organization*

A firm sets its organizational routines to operate the organization efficiently. Stable patterns of behavior form cooperative systems with standard procedures and defined roles. The organization's value criteria and path dependencies influence routines. As a firm's organizational routines become more sophisticated, inertia forces increase; hence, there is a trade-off between efficiency and effectiveness, as shown in Fig. 3.2.

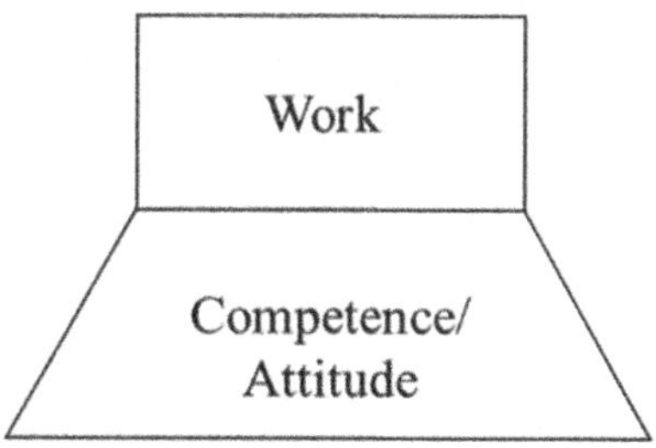

Fig. 3.1 People

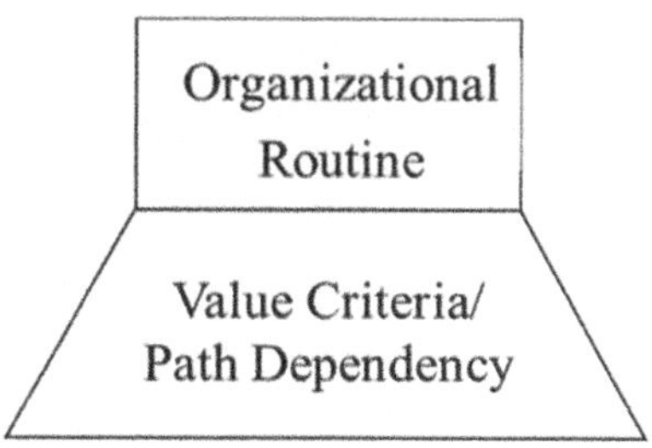

Fig. 3.2 Organization

3.3.3 Business Process

A business process is a mechanism for producing the products and services for customers; it is a collection of business structures and relationships. In general, the flow of business starts from receiving an order, through production, to shipment (business flow). Any firm's business process requires efficiency. To enhance efficiency, mission-critical systems can be introduced to centralize data management in line with business flow. Upgrading business process systems can be a source of a firm's competitive advantage, as shown in Fig. 3.3.

The arrows in the figure show the equilibrium state of "people," "organization," and "business processes," respectively, and the state among the three. In this chapter, equilibrium is defined as a state in which the components are respectively in a stable state, and the three are entirely connected in a stable state.

The work done by "people" cannot make a stable contribution unless it is matched to their competences and attitudes. In addition, the organizational routines of "organization" will not function unless they are in line with the value criteria of the business organization and the so-called path dependencies that the firm has inherited. The same applies to "business processes": Mission-critical systems that do not satisfy the business flow will have little utility value (Fig. 3.4).

Fig. 3.3 Business process

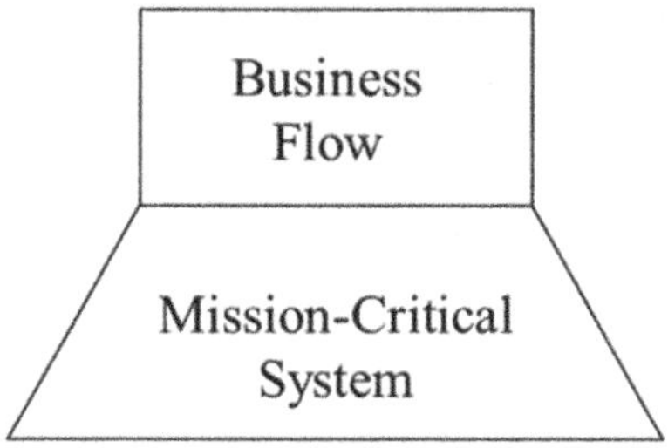

Fig. 3.4 Firms structure with equilibrium

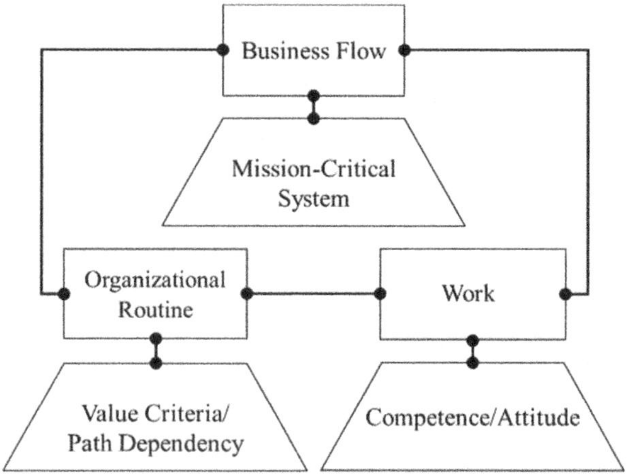

3.4 Dynamism of Dynamic Equilibrium

As a business organization receives daily orders from customers, the equilibrium state can fluctuate from a steady to an unstable state if orders greatly exceed the average processing ability. An unusual amount of workflow to meet a delivery date makes the responsible departments busier and increases the load on the responsible employees. As a result, there is an increased risk of defective products, delivery delays, and erroneous shipments.

If the firm cannot adjust production plans or delivery dates to absorb such disruptions and keep "business processes" in equilibrium, a mutual complementary relationship will be mobilized between "people" and "organization." "People" tend to respond by working overtime or by being supported by members of other departments within the constraints of the organizational routines.

The "business processes" can be disrupted by both the quantitative deviation, due to increased orders, and the qualitative deviation, induced by shorter deliveries or changes in product specifications in the course of production. If these disruptions cannot be absorbed, "people" and "organization" will support each other to maintain a state of equilibrium throughout the organization. This is the dynamism of dynamic equilibrium.

3.5 Characteristics of Dynamism of Dynamic Equilibrium

The essence of dynamic equilibrium is a firm's coordination activities to stabilize an unstable equilibrium. Coordinating activities are also a place of destruction (decomposition) and creation (generation). For example, to respond to unexpected orders from key customers, coordination activities are required at various steps to ensure that the entire operation can proceed smoothly, even after the production plan is modified. In addition, when an employee takes a long leave owing to illness, it is necessary to assign a different person or create a special workflow for that period. In this way, coordination activities create new and stable relationships that break down the existing unstable relationships due to a loss of equilibrium. Coordination activities are always executed with consciousness and the pursuit of an equilibrium state; however, there are limitations based on flexibility.

Organizational routines usually include embedded, rather than stipulated, tacit rules (such as how obtain an internal approval or a method of adjustment/coordination). Thus, deviation activities in the organizational routines can be confounded within acceptable limits. For example, if a manager's leadership causes dysfunction in the flow of work, members without a right of command cannot intervene and provide instructions to correct the problem.

However, coordination activities create an opportunity to reflect on the organizational routines, the business processes and interpersonal relationships that were in a state of equilibrium in the past as well as their respective deep layers. Accordingly, coordination activities can generate a starting point to develop a more evolved equilibrium state. In the previous example, if the manager's lack of leadership is not due to his management skills, but rather to a lack of information, then the focus shifts to the current mission-critical system in the business process. If there is a problem with the seniority-based promotion system, this will lead to a review of the "value criteria and path dependencies."

Otsuki argued that if organizational routines restrict activities, they might need to be reviewed (Otsuki, 2010).

Dynamic equilibrium contains a "seed" to develop the next state of equilibrium, which creates a high dynamism of dynamic equilibrium. Referring to chemical substances in science, a dynamic equilibrium is characterized as "sequential reactions" rather than "chain reactions."

3.6 Strategic Dynamic Equilibrium

A business organization may be disrupted by the radical revamp of organizational routines in a top-down approach, disregarding the organizational value criteria and path dependencies, or by introducing a benchmark-based enterprise resources planning system. Such changes break the current equilibrium and forcibly create a new equilibrium in a discontinuous manner.

The coordination activities undertaken to create a new equilibrium will transform the business organization incrementally, but steadily.

Hence, a "strategic dynamic equilibrium" is the task of embedding intentions in dynamic equilibrium with the aim to build a desirable business organization. For example, to revitalize a sales division, the strategic dynamic equilibrium (as shown in Fig. 3.5) could be considered assuming the seniority-based personnel evaluation system is replaced by a performance-based system to promote the "value criteria which enables employees to improve and develop themselves through a friendly competition."

In the site, information sharing and middle managers' competences are the keys to successful incremental transformation. Through sharing information, problems or failures should be sensed in advance and coordination activities can be started early. The data accumulated in this mission-critical system can also be used effectively for coordination activities.

Middle managers' competences refers not only to their conducting coordination activities with strategic dynamic equilibrium in mind, but also their ability to present to management, unflinchingly, the potential problems that lie in the equilibrium among the current organization, people, and business processes. Thus, they are required to take the role of a self-reflective practitioner.

In turn, it is important for upper management to listen earnestly to middle managers' opinions about what is happening at the site, to avoid missing opportunities to improve current organizational routines, people, and the deep business process layer. In this process, the management and employees must be united. There is no

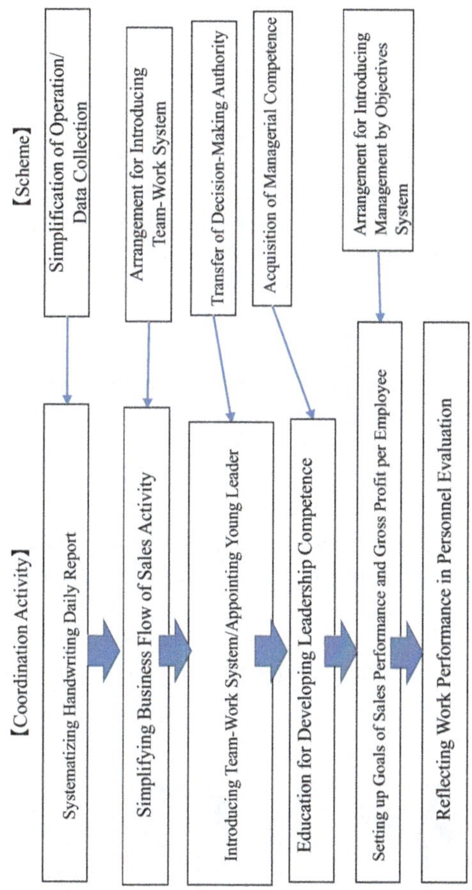

Fig. 3.5 Coordination activity with scheme

doubt that the management philosophy can direct a firm to correct its organizational transformation.

3.7 Conclusion

This chapter models organizational transformation based on the concept of dynamic equilibrium in life science. External factors force the flow of daily business operation to fluctuate, and the equilibrium becomes imbalanced on each occasions. However, the dynamic equilibrium creates a new equilibrium state. In this chapter, the essence of dynamic equilibrium in a business organization is defined as coordination activities. Strategic coordination activities can realize an intended organizational transformation incrementally, but steadily.

Figure 3.6 shows a work created by the artist M.C. Escher, using the regular partition technique (see Escher, 1989), that depicts the transformation of birds into fish from right to left along the time axis. Escher's message is that "even though all sorts of things in the world appear to be changing discontinuously, the essence of 'continuous change' based on one specific pattern" is embedded in the work.

In this respect, a similar pattern applies to organizational transformation: each incremental change, even each minute, will add up to a great change over the course of time. Management must sense small changes, instill effective measures against major environmental changes, and incorporate changes into the daily operations. The proposed model in this chapter designates the appropriate direction of "generation" after the "decomposition" of equilibrium.

Management must make tenacious efforts to clarify the management philosophy and vision until the philosophy and vision are accepted as a code of behavior for all employees in their daily work.

With the globalization of the economy and the rapid advancement of technological innovation, many managers have high expectations for the dynamic capabilities theory.

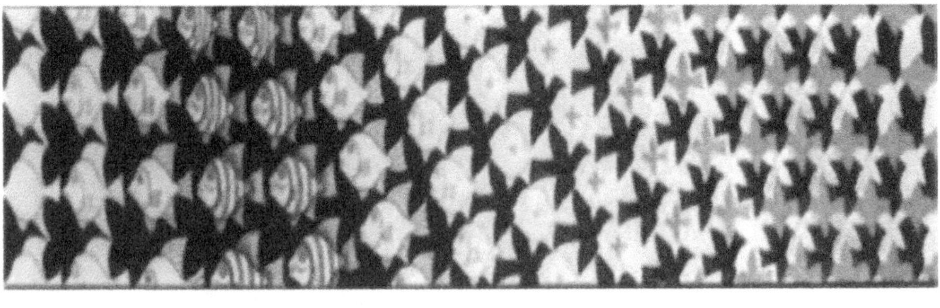

Fig. 3.6 M.C. Escher, metamorphosis (transformation)

This chapter is a challenge because the theory of dynamic capabilities, argued by many researchers, tends to be rather conceptual and comprehensive, while there is little study concerning the theoretical development and practical contributions of procedures and strategies for organizational transformation. In the future, the author intends to tighten the proposed model and present further practical proposals through empirical research.

References

Escher, M. C. (1989) *Escher on Escher: Exploring the infinite* (Japanese-language, I. Sakane (1994), Trans., Asahi Shimbun Publications Inc).

Fukuoka, S. (2009). *Between organic matters and inorganic matters*. Kodansha Ltd.

Fukuoka, S. (2011). *Dynamic equilibrium 2*. Kirakusha Inc.

Fukuoka, S. (2017). *Dynamic equilibrium 3*. Kirakusha Inc.

Huang, Y. (2011). *A challenge and possibility on the theory of dynamic capability* (Departmental Bulletin Paper, Vol. 3, pp. 29–42). Waseda University Graduate School of Commerce.

Otsuki, H. (2010). Organizational transformation and dynamism of organizational routines. *The Waseda Commercial Review, 425*.

Teece, D. J. (2007). Explicating dynamic capabilities: The nature and micro foundations of (sustainable) enterprise performance. *Strategic Management Journal, 28*, 1319–1350.

Teece, D. J. (2012). Dynamic capabilities: Routines versus entrepreneurial action. *Journal of Management Studies, 49*, 1395–1401.

Teece, D. J., Pisano, G., & Shuen, A. (1997). Dynamic capabilities and strategic management. *Strategic Management Journal, 18*(7), 509–533.

Zollo, M., & Winter, S. G. (2002). Deliberate learning and the evolution of dynamic capabilities. *Organization Science, 13*(3), 339–351.

Chapter 4
A Process for a Conceptual Design and Its Simulation Toward New Business Model Creation

Kazuki Yoshida

Abstract A process for a conceptual design and its simulation toward new business model creation is proposed. In this process, a SWOT analysis on Business Model Canvas (BMC) is done to clarify strength and weakness of the company, then Business Model Pattern (BMP) is chosen to formulate its creation policy. According to the policy, Action Framework (AF) is applied to decide the measures for realizing the new model, and the migration process to the new model through those measures is simulated using System Dynamics (SD) and Monte Carlo Simulation (MC) in order to verify its feasibility, profitability, and growth.

The novel points in this proposal are described as follows.

1. A series of tasks from conceptual design to verification of a new business model is defined as a process.
2. Incorporating BMP as a template for new business models and formulating policies
3. For each block of the BMC, the method of building SD models is standardized.
4. For each block of the BMC, the method of building the extended part of the SD model is standardized according to the results of AF application.
5. For each BMP, parts that can be formalized as SD model are made and applied to extend the SD model.
6. When multiple BMPs are applied, the tasks to be performed in conceptual design and the rules to be followed in verification are prescribed.

Furthermore, this chapter is written on the basis of (Yoshida, Japan Society for Management Information 29, 2021) and (Yoshida, Journal of the Japan Society for Management Information 30, 2022).

Keywords Business model · Business model canvas · Business model pattern · System dynamics · Monte Carlo simulation

K. Yoshida (✉)
National Institute of Informatics, Tokyo, Japan

K. Kijima et al. (eds.), *Systems Research II*, Translational Systems Sciences 27,
https://doi.org/10.1007/978-981-16-9941-2_4

4.1 Introduction

There is a growing awareness of the need to create new business models. According to a survey, companies that innovate their business models are more profitable than companies that innovate their products and processes (Gassmann et al., 2016). In addition, in many cases, the decline of large traditional companies with a long history and good name is considered to be due to their inability to adapt their business models to the changing environment. Currently, many companies are facing a similar situation (Ministry of Economy, Trade and Industry, 2018).

As a result of their research, Gassmann et al. extracted 55 repeating patterns that form the core of business models that have proven successful in the past, and stated that these patterns can be used as a template for creating new business models. They also state that more than 90% of the creation of new business models is a combination or adaptation of these patterns (Gassmann et al., 2016).

Therefore, in this chapter, based on the business model design evaluation method (Minato, 2013) that integrates the business model canvas (Osterwalder & Pignol, 2012) (hereafter referred to as BMC) and system dynamics (hereafter referred to as SD), we investigate the conceptual design of a new business model at the conceptual stage by combining the SWOT analysis and action framework ((Osterwalder & Pignol, 2012) (hereafter referred to as AF), and business model patterns (hereafter referred to as BMP) on BMC, then we propose a process to design a new business model at the conceptual stage and to verify its feasibility, profitability, and growth potential.

Specifically, in this process, BMC is adopted as the architecture of the business model, and based on the results of SWOT analysis of the current business model, the policy for creating a new business model is formulated by applying BMP. Then, for each block of the BMC, we write down the measures to be implemented in accordance with the AF. From this, the SD model is constructed and simulations are performed to verify the feasibility, profitability, and growth potential of the new business model.

Compared with the business model design evaluation method based on the integration of BMC and SD proposed in Minato (2013), the proposal in this chapter adds the following new elements.

1. A series of tasks from conceptual design to verification of a new business model is defined as a process.
2. Incorporating BMP as a template for new business models and formulating policies.
3. For each block of the BMC, the method of building SD models is standardized.
4. For each block of the BMC, the method of building the extended part of the SD model is standardized according to the results of AF application.
5. For each BMP, parts that can be formalized as SD model are made and applied to extend the SD model.
6. When multiple BMPs are applied, the tasks to be performed in conceptual design and the rules to be followed in verification are prescribed.

The structure of this chapter is as follows.

In Sect. 4.2, we explain the conceptual design process of a new business model. 2 and 6 above are explained in this process.

In Sect. 4.3, we explain the verification process of the new business model. The above 3–6 are explained in this process.

In Sect. 4.4, we summarize the measures adopted in the above process to obtain the same verification results uniformly and their issues, as well as the other issues that became apparent while applying and implementing multiple patterns.

In Sect. 4.5, we survey the results of domestic and international researches related to the research theme of this chapter.

In Sect. 4.6, we summarize this chapter.

In order to explain the proposed process, we use the example of "receiving a home delivery package at a convenience store (hereafter referred to as CVS)."

4.2 Conceptual Design Process for New Business Models

In this section, we first identify the strengths, weaknesses, opportunities, and threats of the current business model for each block of the BMC. Then, we formulate policies to deal with these issues, referring to the BMPs that serve as the template for the new business model. Then, based on the policies, we identify the measures to be implemented in each block. Furthermore, when multiple BMPs are provided, we identify the interactions between them. In the following sections, we will explain these methods, showing the results of application using examples.

4.2.1 SWOT Analysis

In this conceptual design process, SWOT analysis is combined with BMC to conduct a detailed SWOT analysis (Osterwalder & Pignol, 2012) focusing on each of the nine blocks of BMC (hereafter referred to as "SWOT on BMC"). The purpose of this analysis is to make it easier to link the results of the analysis to the subsequent formulation of policies and measures, rather than focusing on the entire business model. This detailed SWOT analysis provides a snapshot of the current strengths and weaknesses of the business model, as well as information on opportunities and threats that should be considered when formulating policies for the creation of a new business model.

An example of SWOT on BMC for the home delivery industry is shown in Fig. 4.1. For example, looking at the customer segment (CS), it is an opportunity that the number of customers who use home delivery services is increasing due to the spread of online shopping. On the other hand, such customers often have delivery time constraints, and in many cases, they are unable to receive their packages even at the specified time and have to have them redelivered. This means that how the delivery industry copes with the increase in costs associated with such constraints will become a threat to the operation of delivery services.

4.2.2 Formulating a Policy

In (Osterwalder & Pignol, 2012), a tool called AF is proposed to create a new market space without competition through value innovation. It consists of the following four questions

1. Which elements should be removed that are taken for granted in the industry?
2. Which elements should be reduced to below industry standards?
3. Which elements should be increased over the industry standard?
4. What elements should be added that the industry has not previously provided?

By considering the answers to these questions, we can reduce costs by reducing or eliminating services and functions that have no value, and increase or add functions and services that provide high value without increasing costs (Osterwalder & Pignol, 2012).

In formulating policies for a new business model, based on the strengths and weaknesses of the business model identified through SWOT on BMC in Sect. 4.2.1, and taking into account opportunities and threats, we decide how to deal with them by applying the AF described above. Basically, there are three types of policies that can be considered.

1. To increase the volume of factors that are strengths in order to further increase the difference from competitors
2. Decide not to compete with other companies on weaknesses and reduce/eliminate those weaknesses
3. Increase/add weak elements in order to be able to compete with competitors

After recognizing the strengths and weaknesses of the current business model, the BMP provides a template for what kind of new business model can be considered when increasing, reducing, adding, or removing those elements.

In order to select the appropriate BMP for your policy development from among the 55 BMPs, refer to the What (what to offer to customers), Who (who is the target customer), Why (why it will be profitable), and How (how to generate a value proposition) described in the Basic Pattern section.

For example, in Self-service pattern, one of the 55 BMPs, it is noted that the business model of this BMP has a great potential for cost reduction (Why). This means that

- Excessive costs in certain parts of the current business model exist.
- We want to change this part of the process to be implemented by the customer (how).
- This will save money and time for the customer (what).

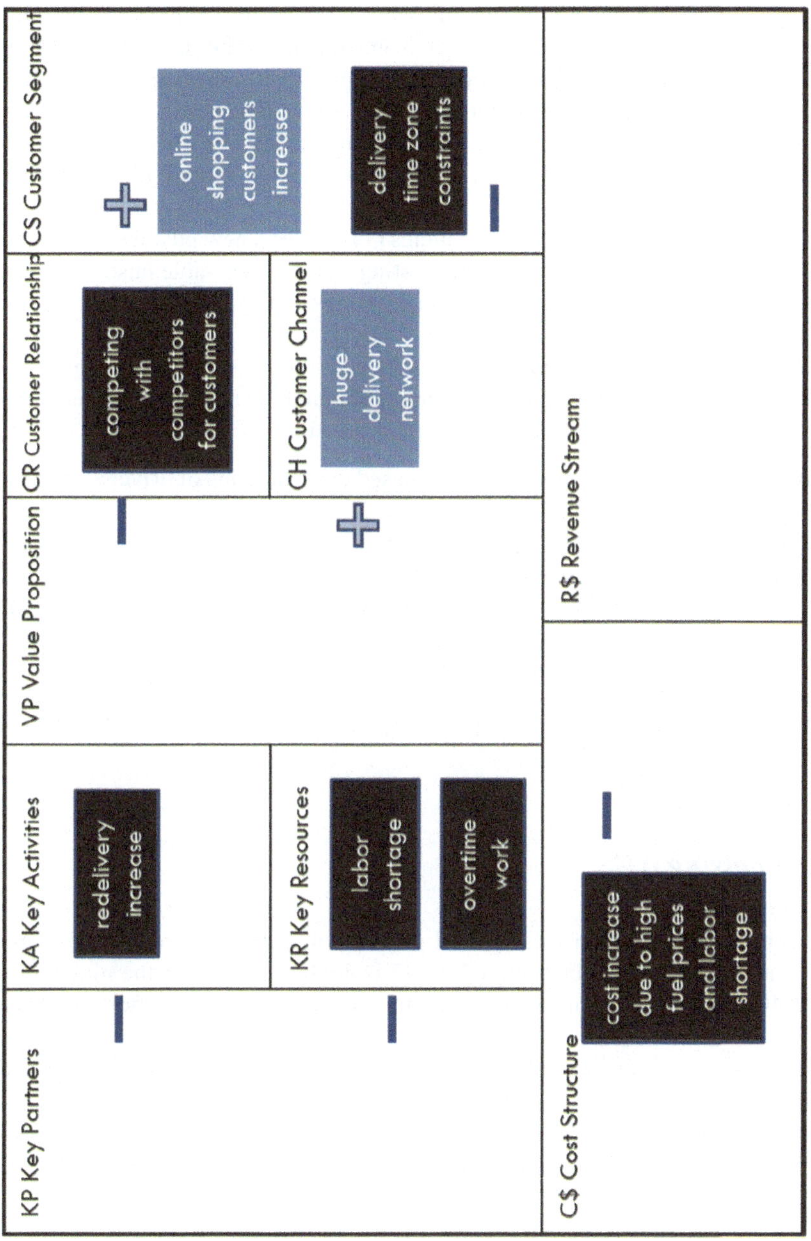

Fig. 4.1 Example of SWOT on BMC for the home delivery industry

These points are the key to determine whether it is desirable to change the business model by applying this BMP or not.

In addition to the application of new BMPs, the replacement of BMPs that are already applied in the business model with other BMPs is also considered in policy formulation. Among the 55 BMPs, there are some that have alternative relationships with each other as solutions to specific business problems. For example, Flat Rate pattern, Pay Per Use pattern, Subscription pattern, Add-on pattern, and so on can be classified as solutions to the problem of the revenue model. For BMPs belonging to the same category, if a BMP is already applied in a business model, it is possible to replace it with another BMP.

In this way, in accordance with the policies (1)–(3) for a new business model, we will try to create a new business model by listing as many desirable business model options as we can think of, referring to the 55 BMPs, and selecting an appropriate one among them.

We consider the case of applying Self-service pattern as a template for a new business model to appropriately deal with the problem of increasing redelivery costs in the home delivery industry. The above points are summarized as follows.

- The increase in redelivery is causing increased costs in terms of driver's overtime work and trucks' fuel.
- Instead of delivering packages to the customer's home, let the customer come to the nearest CVS to pick them up.
- Customers will be freed from the constraint of having to be at home during the delivery time, and will be able to go to CVS and pick up their packages at a convenient time, in addition to other purposes.

From these points, we can conclude that it is desirable to apply Self-service pattern.

4.2.3 Applying AF

In this conceptual design process, based on the policy for the creation of a new business model formulated in Sect. 4.2.2, AF is applied to each of the nine blocks of the BMC to identify the measures to be implemented there (hereafter referred to as "AF on BMC").

Figure 4.2 shows an example of the application of Self-service pattern, AF on BMC, based on the customer's receipt of a package at CVS. This means, for example, that Customer Channel (CH) presents a new option for customers who use home delivery to pick up packages at CVS, while Key Partner (KP) will outsource the storage of packages to CVS so that they can pick up packages at CVS.

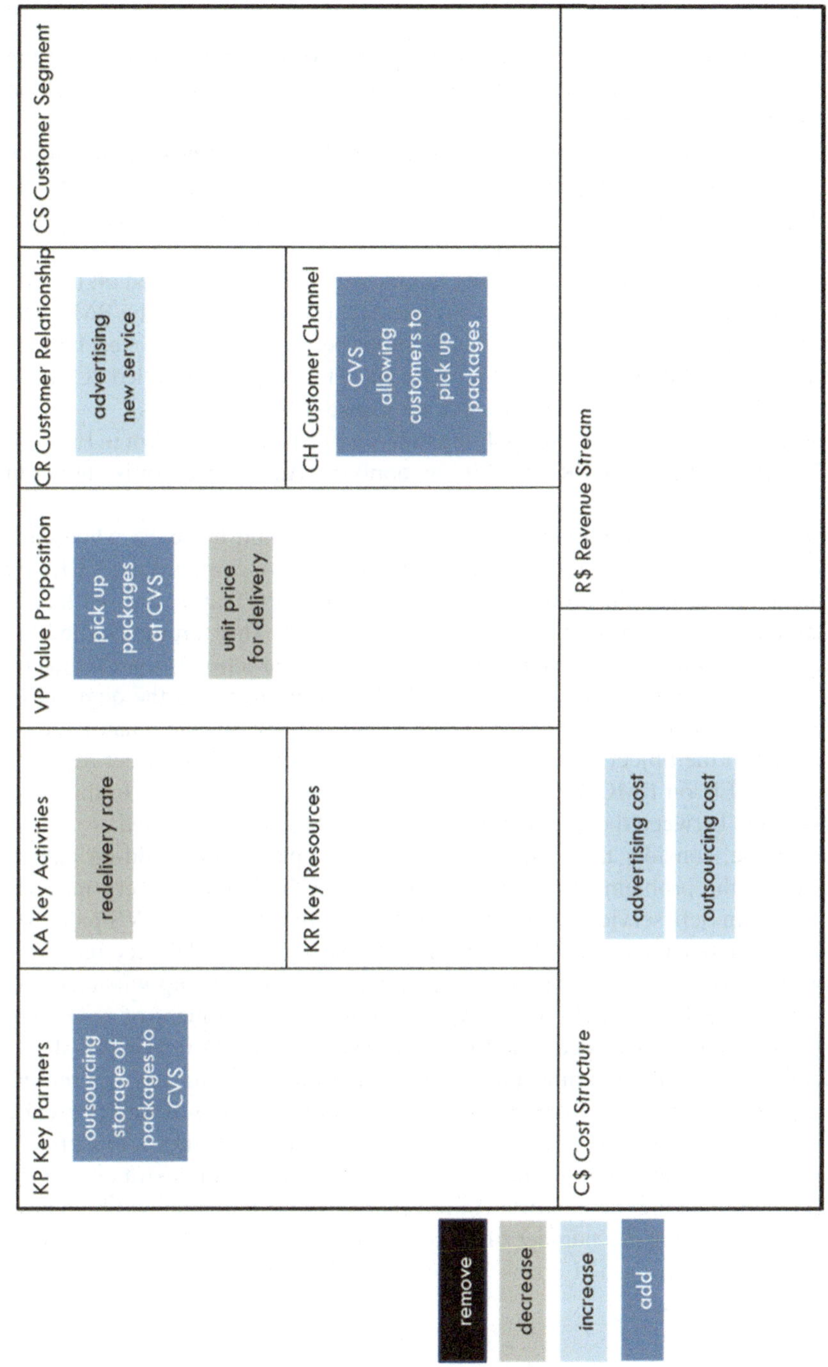

Fig. 4.2 Example of AF on BMC based on the application of Self-service pattern to the home delivery industry

4.2.4 *Identifying Interactions Between Multiple BMPs*

To address the problem of increased redelivery, the AF on BMC for reducing the redelivery rate by applying Self-service pattern and allowing customers to pick up their packages at CVS is shown in Fig. 4.2 in the previous section.

On the other hand, the AF on BMC is shown in Fig. 4.3 when Add-on pattern is applied, the activity of redelivery is extracted from the delivery service, and the service is provided with a fee structure in which the basic charge for delivery and a separate charge for redelivery are added.

Then, for each BMP, by implementing the measures organized based on the AF, we replace the term with one that describes the state of each block of the BMC (the result state), and describe the cause–effect relations between the states of each block again on a single BMC in the form of arrow lines. This diagram is called "CER on BMC by applying ○○ pattern" (CER stands for Cause–Effect Relations).

For example, CER on BMC by applying Self-service pattern is shown in Fig. 4.4.

On the other hand, the CER on BMC by applying Add-on pattern is shown in Fig. 4.5.

Interactions between multiple BMPs are those that could not exist when each BMP was applied alone. To identify these interactions, we compare the CERs on BMCs made for each application of BMPs, identify the parts that interact with each other, and extract only those parts, as shown in Fig. 4.6. This figure is called the "Interaction diagram of ○○ pattern and △△ pattern". The action is represented by a thick dashed arrow pointing at the target. If the action increases the quantity of the object, a "+" sign is added to the object, and if it decreases the quantity, a "-" sign is added to the object. In this figure, the BMP-specific cause–effect relations identified by CER on BMC are also indicated by thin dashed arrows. In this way, the relationship between the interaction and the causality is also clarified.

For example, consider the case where Self-service pattern and Add-on pattern are applied to the problem of increased redelivery in the home delivery industry (see Fig. 4.7). In Self-service pattern, the unit price of the charge for CVS pickup is reduced compared to the unit price of the charge for normal home delivery, but if this reduction becomes larger, the number of customers of CVS pickup will increase, and as an interaction to Add-on pattern, the number of customers of redelivery with a separate charge will decrease (the thick dashed arrow 1) in Fig. 4.6 shows this effect). If the total amount of charges including the extra charge is increased, the number of customers of CVS pickup is expected to increase as an interaction with Self-service pattern (the thick dashed arrow 2) in Fig. 4.6 shows this effect). In addition, increasing the total amount of the fee, including the extra charge for redelivery, and reducing the unit price of the charge for CVS pickup will both have the effect of increasing the number of customers of CVS pickup (the thick dashed arrow 2) and the thin dashed arrow 3) are directed to the "transition+" into the "CVS pickup customer").

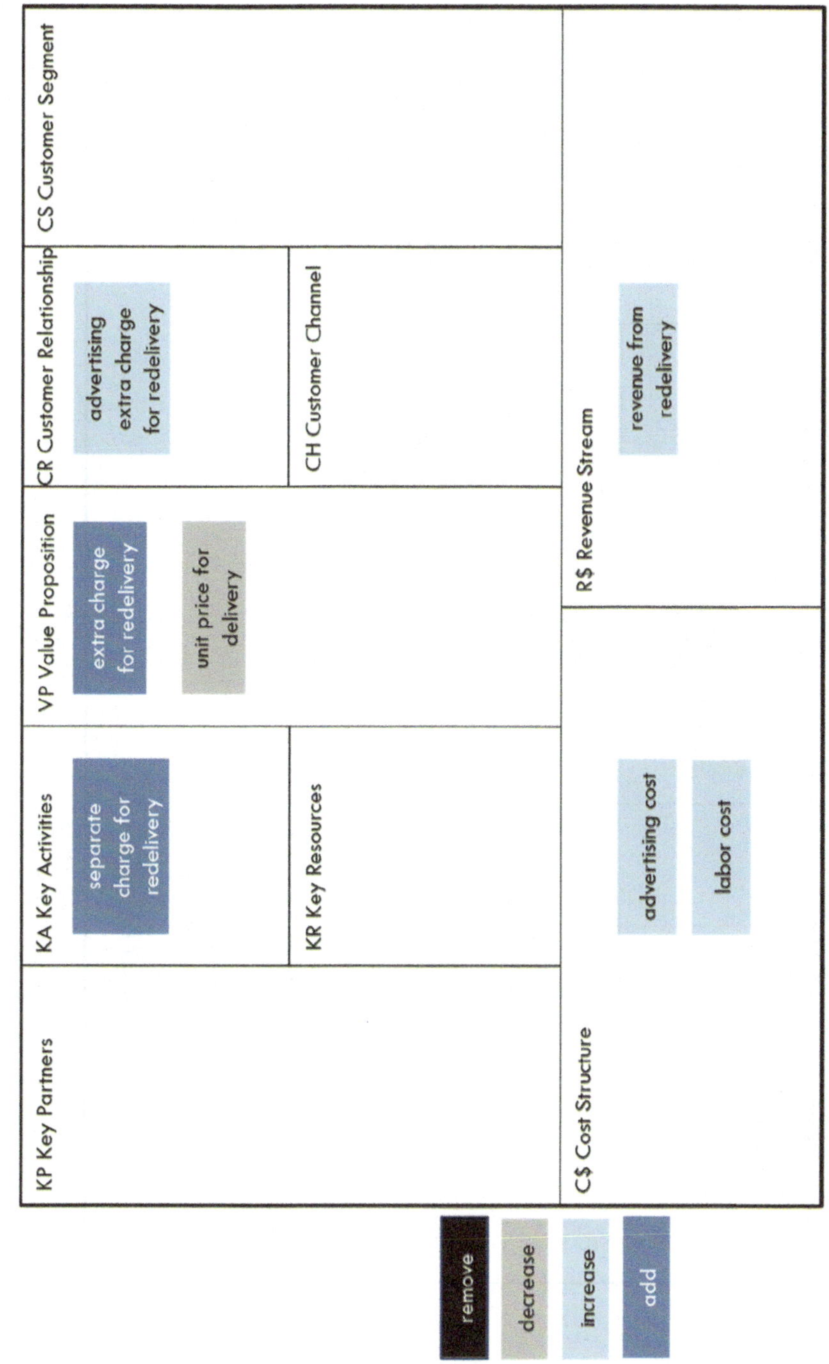

Fig. 4.3 Example of AF on BMC based on the application of Add-on pattern to the home delivery industry

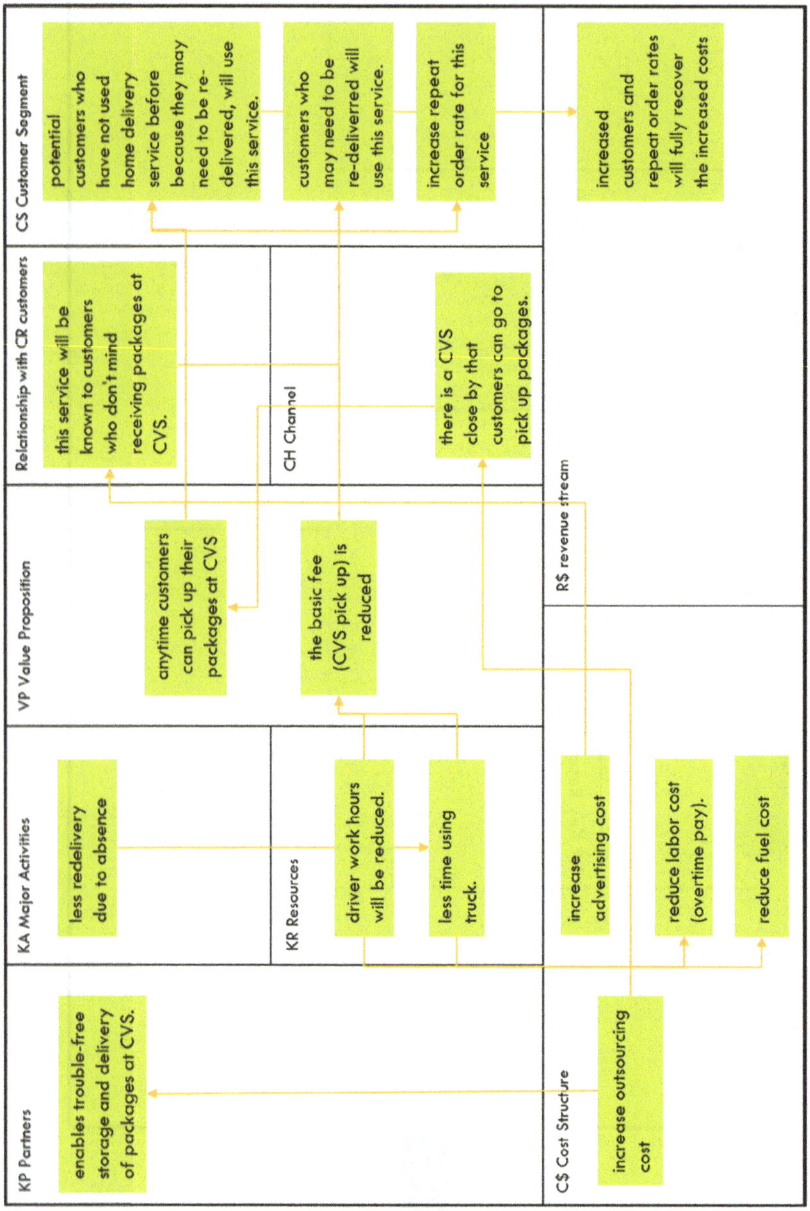

Fig. 4.4 An example of CER on BMC by applying Self-service pattern to the home delivery industry

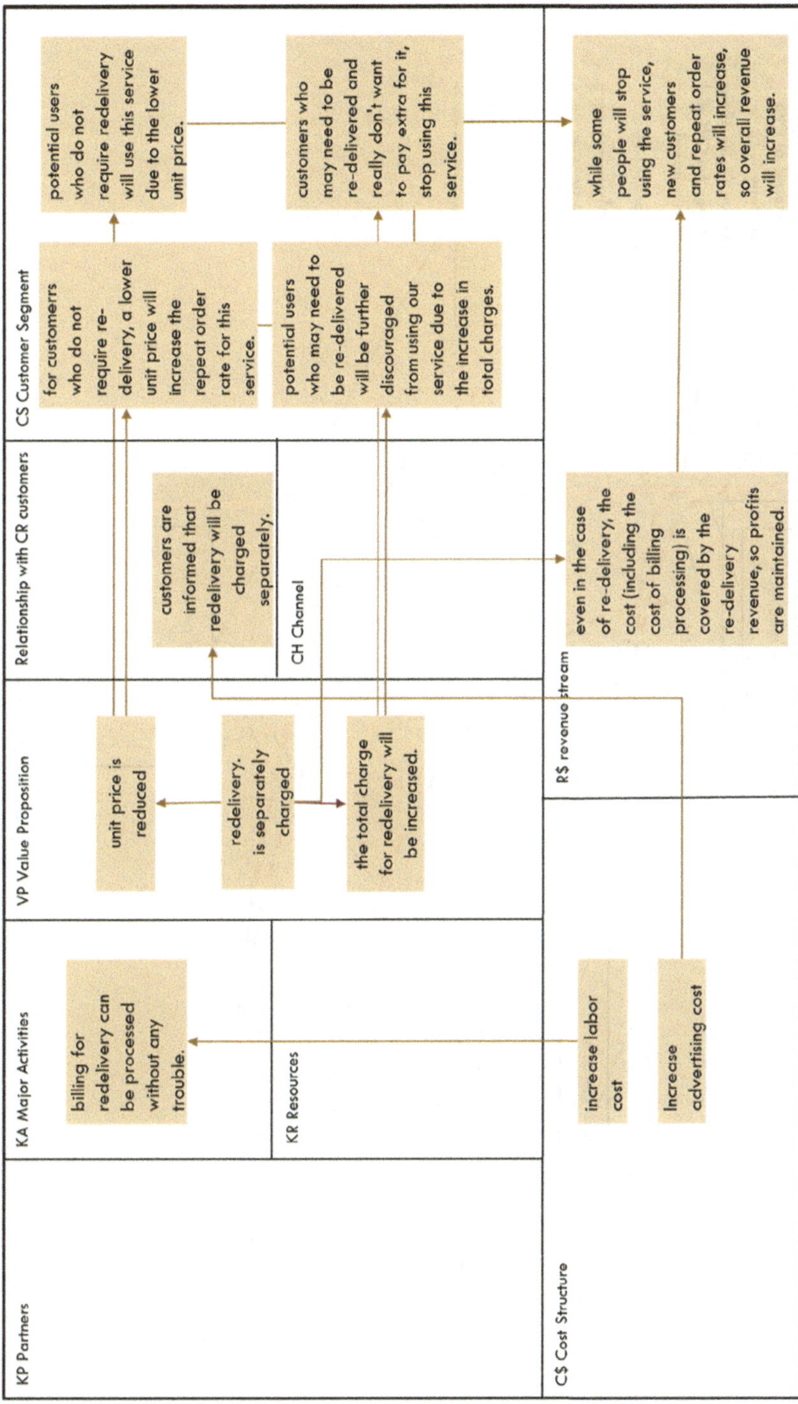

Fig. 4.5 Example of CER on BMC by applying Add-on pattern to the home delivery industry

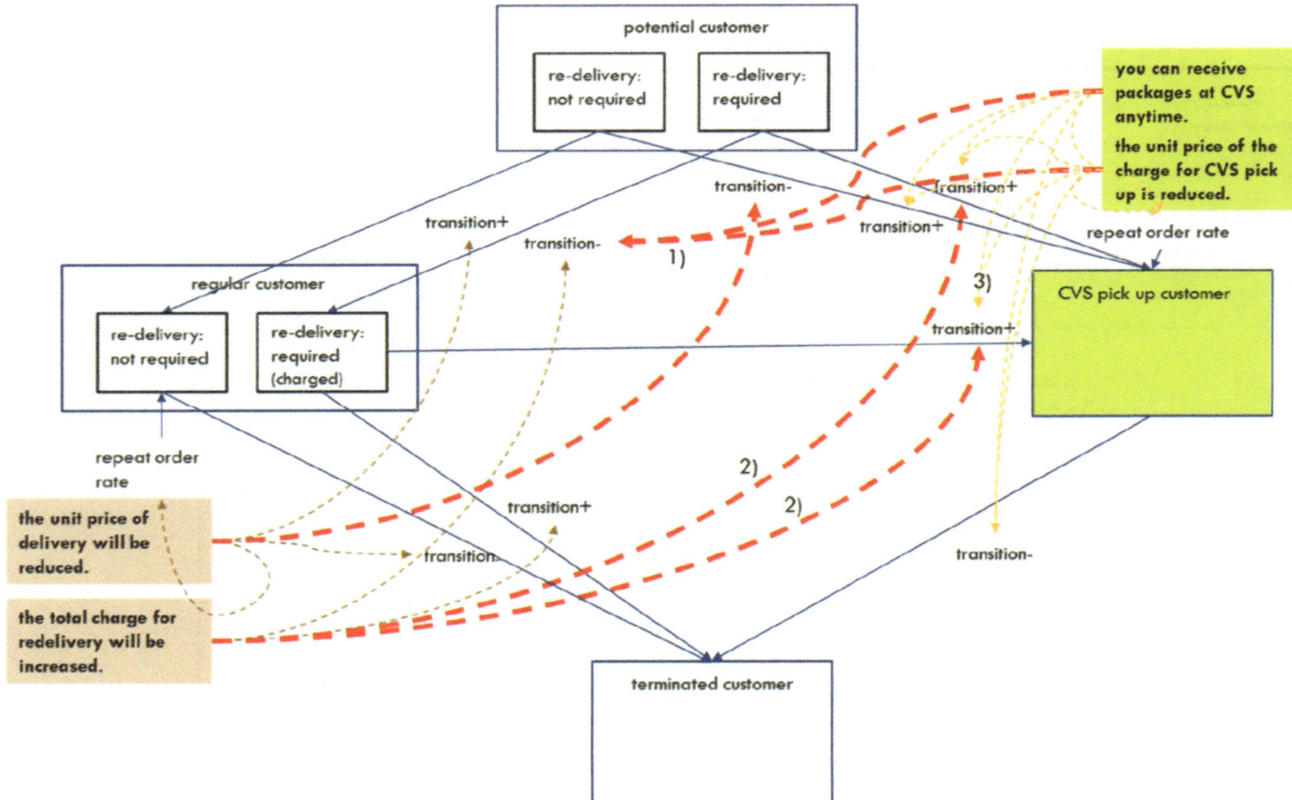

Fig. 4.6 Example of interaction diagram between Self-service pattern and Add-on pattern

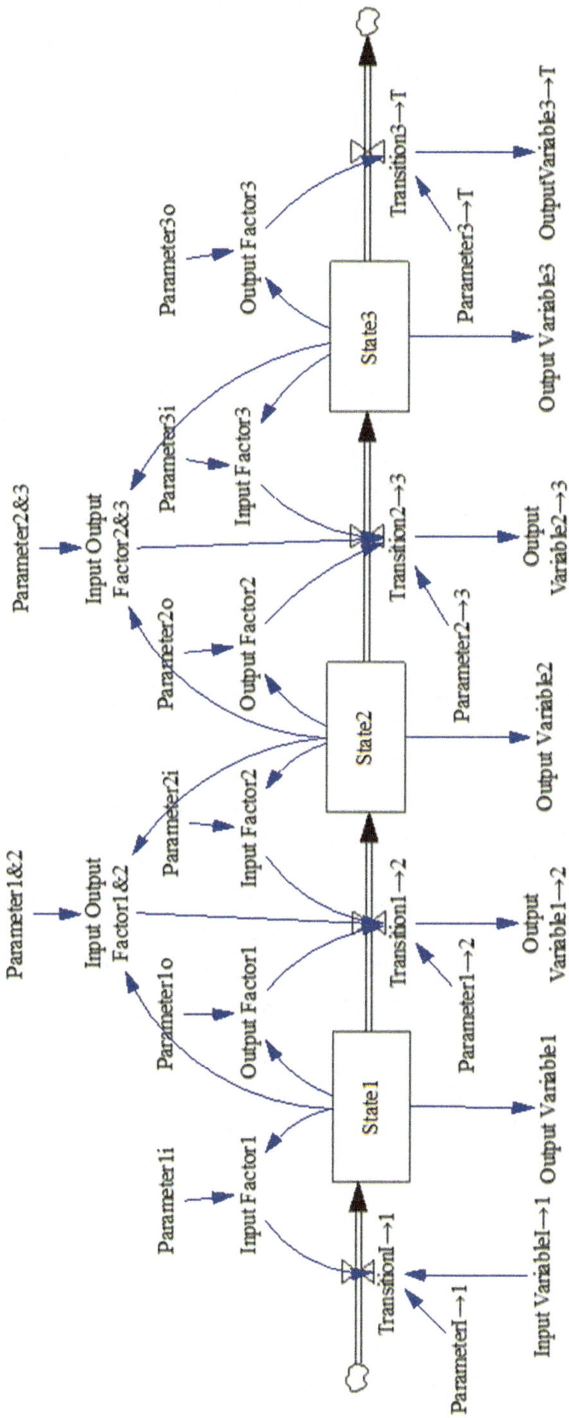

Fig. 4.7 SD model in KA/CR/CH block (conceptual diagram) in case of state defined as stock

By the way, in this example, CER on BMC and interaction diagrams were created assuming that the current model already exists and that Self-service and Add-on patterns are newly applied to it. The same task will be performed between the cause–effect relations existing in the current model, or the BMPs already applied, and the BMPs to be newly applied. The activity of creating business models by applying BMPs is to be repeated continuously on a long-term time axis, and for this purpose, it is necessary to always perform this task with the current model at any given time.

4.3 Verification Process for New Business Models

In this section, we first build the SD model that represents the current business model on BMC. On the other hand, the new business model after the implementation of the measures identified in AF on BMC is also built as an SD model and merged into the current SD model. Then, we verify the feasibility, profitability, and growth potential of the new business model by running simulations that include the transfer of customers to the new business model. In the following sections, we will explain these results using examples.

4.3.1 Description of the Current Business Model

Minato (2013) integrated SD into BMC to reproduce the dynamic behavior of people, goods, and money in a time-series manner so that the feasibility, profitability, and growth potential of a business model can be evaluated simultaneously on a single canvas.

In this verification process, we build SD models for the current business models by integrating SD into BMC, following Minato (2013). In this process, from the viewpoint of defining the process, the method of building SD models is standardized and specified for each block of BMC. This formulation is also applied to the case of a new business model.

For each block of KA/CR/CH, we focus on the business process, the "people" and "things" that appear in the process, the organization to which they belong, and the state that changes as the business process progresses. We define the state of the organization to which they belong and the state that changes as the business process progresses as a stock. Then, the transitions of the states of "people" and "things" and the flows among organizations are defined as flows (see Fig. 4.7). Here, the number of "people" and the quantity of "things" are stored in stocks and variables. In the case of business models, the unit of time in SD models is usually "week" or "month" at the minimum. Therefore, we do not elaborate on the business process, but macroscopically treat the number of people and quantity per week/month in stock units, and build the model.

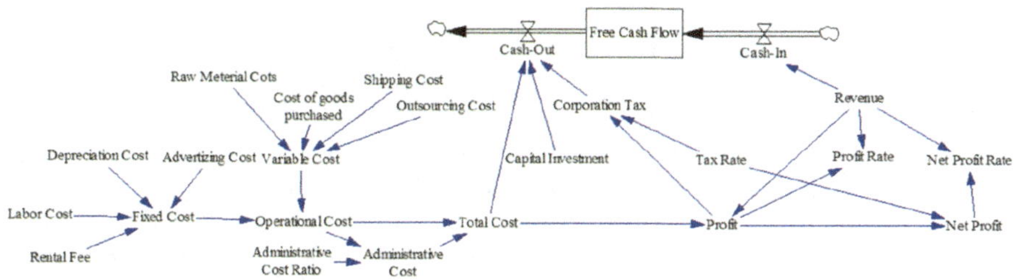

Fig. 4.8 SD model in the C$/R$ block

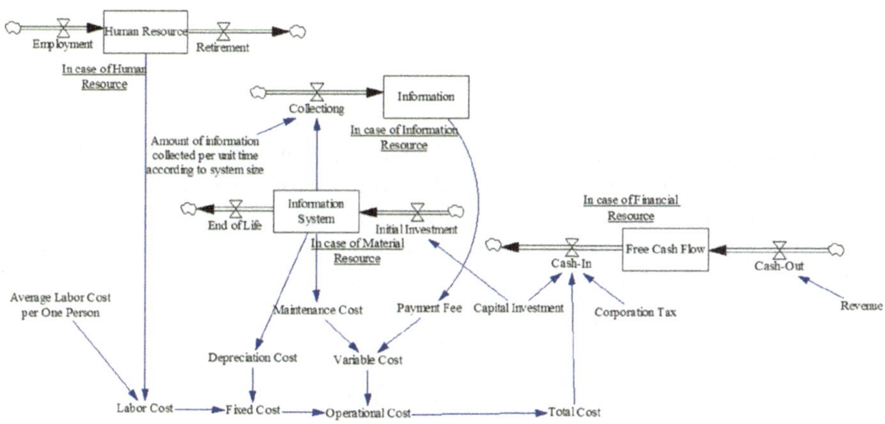

Fig. 4.9 SD model in KR block

For each block of C/R$, since the final purpose is to verify the feasibility, profitability, and growth potential of the new business model, following (Minato, 2013), we define the flow of "money" or "things converted into money" among the various indicators of the income statement as variables (see Fig. 4.8). Therefore, it is the amount of money that is accumulated in stocks and variables.

For the KR block, various resources are treated as stocks. All the activities in the KA/CR/CH blocks are modeled as consuming the various resource stocks in the KR block and reflecting them in the index of the C$ block (see Fig. 4.9). However, the stock of "free cash flow," which represents "cash" as one of the resources, is positioned across the KR/C$/R$ blocks in the sense that it belongs to all of them.

For the CS block, we define the SD model by expressing customer segments as stocks and customer transition between segments as flows, respectively, referring to the SD model of the BASS model (Sakai & Kawai, 2005), which is a diffusion model for new products/services (see Fig. 4.10).

The SD model of the current business model of the home delivery industry is shown in Fig. 4.11 as the SD model built according to this formulation.

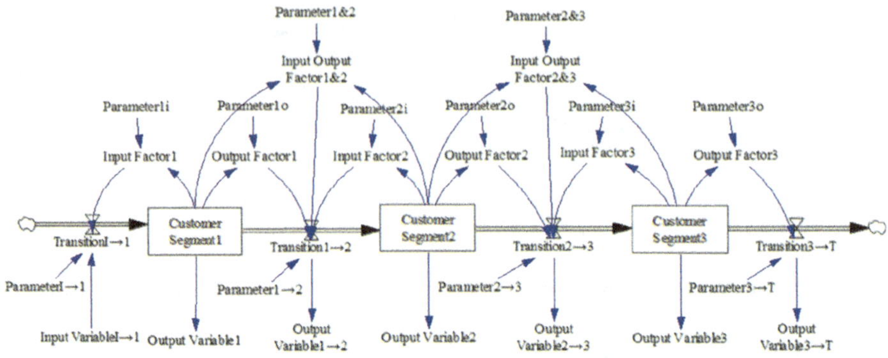

Fig. 4.10 SD model in CS block (conceptual diagram)

This SD model is classified by BMC blocks as shown in Table 4.1.

In the SD model shown in Fig. 4.11, orders from customers are delivered together with their packages, and the orders are managed in the backlog, and the packages start from the distribution center. When the packages are delivered, the orders in the backlog are considered to be completed and are removed from the backlog.

4.3.2 Reflecting the Conceptual Design

To the SD model that represents the current business model, we create a business model that implements the measures identified in Sect. 4.2.3 as an extension of the SD model by adding new elements (stock, flow, variables, and relationship lines). In the CS block, we divide the segments into potential customers, adopters of conventional products/services, and adopters of new products/services based on the new business model, and build the SD model so that we can simulate the implementation of the business model and the resulting transition of customers among the segments.

In this process, the extended part of the SD model is built by modifying the current SD model based on the results of AF on BMC, according to the type of measures to be implemented for each block ("remove," "reduce," "increase, "and "add"). In this verification process, from the viewpoint of process definition, these changes are standardized and specified.

1. "Remove."

   In the specified block, the following two points are eliminated from the business model by removing the stock corresponding to the content of the measure and the flow associated with it, as well as the relation lines drawn from the flow to other elements.

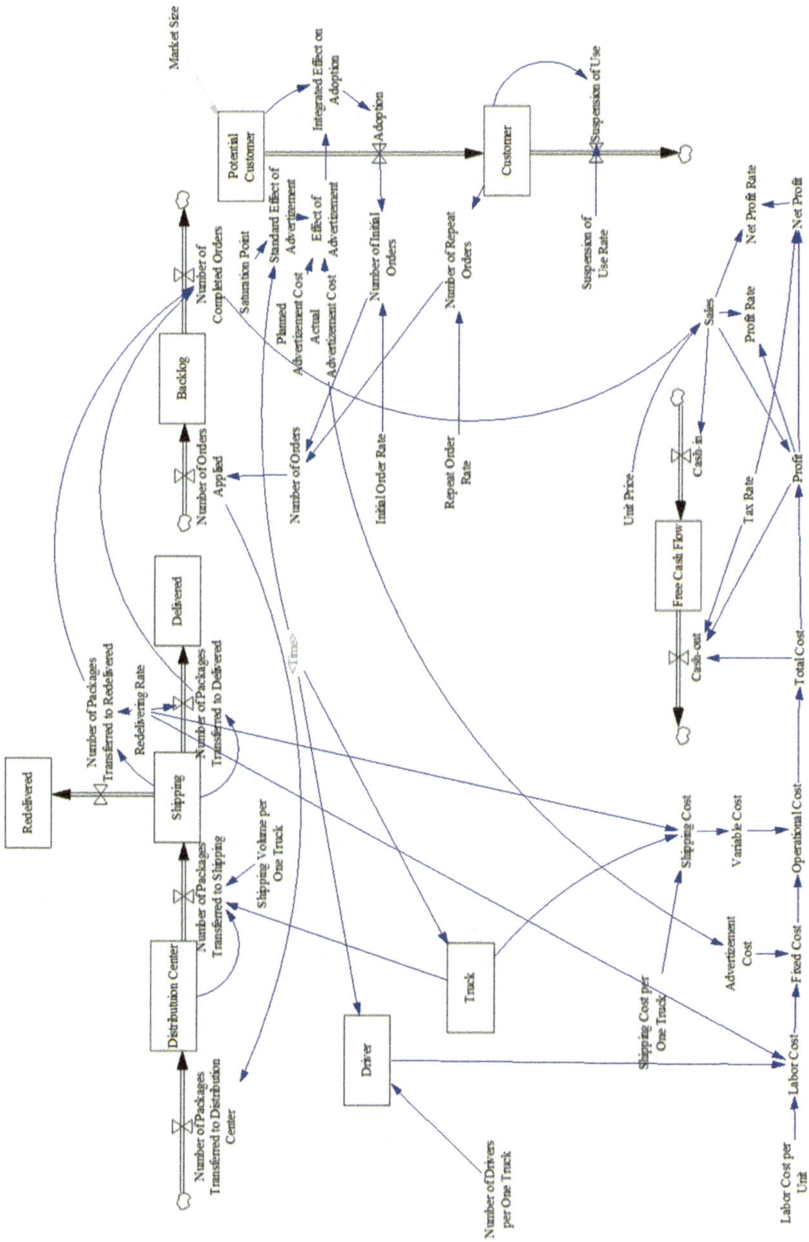

Fig. 4.11 SD model of the current business model of the home delivery industry

Table 4.1 Classification of the SD model (Fig. 4.11) by BMC blocks

Block	Stock	Variable
KA	Distribution center, shipping, delivered, redelivered, backlog	Shipping volume per one truck
KR	Driver, truck, free cash flow	Number of drivers per one truck
CS	Potential customer, customer	Market size, adoption, number of initial orders, number of repeat orders, suspension of use rate
CR		Actual advertisement cost, planned advertisement cost, effect of advertisement, standard effect, saturation point, initial order rate, repeat order rate
VP		Unit Price
C$.	Free cash flow	Shipping cost per one truck, shipping cost, labor cost per unit, labor cost, advertisement cost, fixed cost, variable cost, operational cost, Total cost, tax rate
R$	Free cash flow	Sales, profit, profit margin, net profit, net profit margin

1. Business model effects based on cause–effect relations expressed by relation lines.
2. The impact it had on KR/C$/R$ in terms of resource consumption, cost and income increase/decrease.

2. “Reduce.”
   Within the specified block, implement one of the following three types of changes to the variable corresponding to the content of the measure

   2.1. Reduce the value of the variable connected to the flow (input side) by the relation line to reduce the amount of flow into the stock.
   2.2. Reduce the amount of outflow from the stock by reducing the value of the variable connected to the flow (output side) by the relation line.
   2.3. Reduce the values of variables that affect KR/C$/R$.

3. “Increase.”
   Within the specified block, implement one of the following three types of changes to the variable corresponding to the content of the measure.

   3.1. Increase the value of the variable connected to the flow (input side) by the relation line to increase the amount of flow into the stock.
   3.2. Increase the value of the variable connected to the flow (output side) by the relation line to increase the outflow from the stock.
   3.3. Increase the value of variables that affect KR/C$/R$.

In (2) and (3), in terms of resource consumption, increase or decrease of cost or income, how it affects KR/C$/R$ follows what is defined in the traditional SD model.

4. "add to."

   Within the specified block, a new stock and its associated flow are added according to the content of the measure, and a relation line is drawn from the flow to the other elements to newly define the following two points.

   4.1. Business model effects based on cause–effect relations.
   4.2. Impact on KR/C$/R$ in terms of resource consumption, increase/decrease in costs and revenues.

In this verification process, we reuse the parts of each BMP that can be formalized as SD models.

1. However, if the specific scenario that represents the situation at the time of application becomes clear, it is going to be a submodel that can be incorporated into the SD model by selecting specific elements in the variable part according to the scenario. This part is called the SD model template because the customer selects elements at his/her discretion according to the scenario. Self-service pattern (see Fig. 4.12), Experience Selling pattern, etc. fall under this category.

2. It is possible to make a part in the form of a fixed submodel that does not require customization. This part is called SD model component because it can be applied without customization. Add-on pattern (see Fig. 4.13), Subscription pattern, Freemium pattern, etc. fall under this category.

3. It is not applied to business models as an add-on, but as a foundation for all business models, on which many business models are built. Since it is applied as a foundation, this part is called SD model framework. Two-sided Market pattern (see Fig. 4.14), for example, falls under this category.

4.3.3 Describe the New Business Model

As the SD model built according to the ideas described in the previous sections, a new business model in which Self-service pattern and Add-on pattern are applied step by step to the current business model of the home delivery industry to reduce the ratio of redelivery from the business process of home delivery is shown in Sects. 4.3.3.1 and 4.3.3.2. In this SD model, the redelivery rate, which represents the ratio of redelivery, is multiplied when a customer transitions from being delivered to already delivered in the KA block of the current model, whereas a process without redelivery is defined separately in the same block, and the customers of the new service transit into that process.

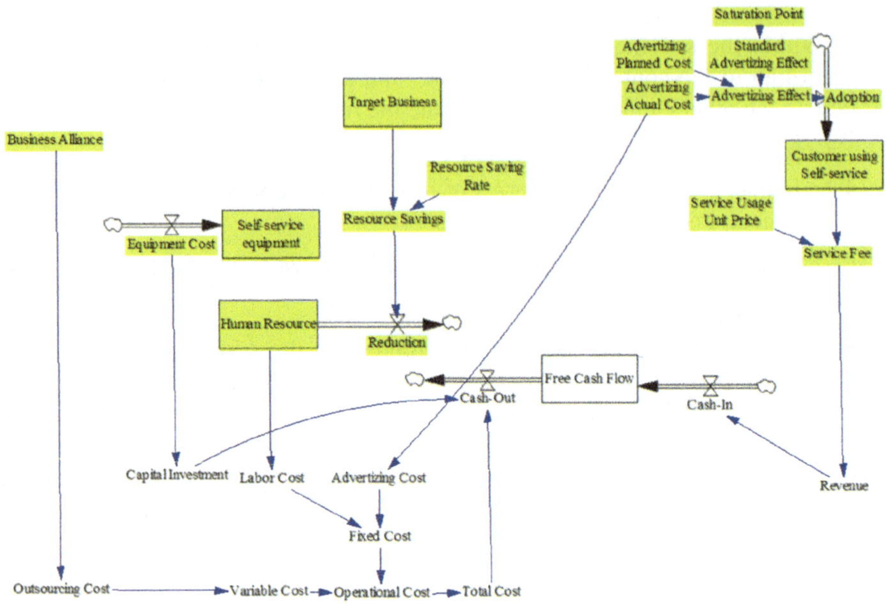

Fig. 4.12 SD model template for Self-service pattern (shaded in yellow). This is an example of investment in CVS (business alliance), which is an element of KP, in order to encourage customers to take action, which is the purpose of Self-service pattern. Another possible example is to invest in lockers (facilities for self-service), which are an element of KR, as in the Amazon Hub locker (Amazon hub, 2021) being introduced by Amazon. In both cases, however, the same point is true: the investment will lead to the creation of a new option for customers, and the customers who adopt the new option will migrate to a new customer segment (self-service customers) within the CS. Therefore, whether the investment is in KP or KR becomes a variable selection factor

4.3.3.1 SD Model when Applying Self-Service Pattern

First, Fig. 4.15 shows the SD model when Self-service pattern is applied to the current business model and packages can be picked up at CVS.

The elements newly added to the SD model of the current business model are shown in red. In addition, the parts where the SD model template of Self-service pattern is applied are shaded in yellow. This includes the cause–effect relations specific to Self-service pattern, which are also indicated by the yellow relation lines based on the results identified by CER on BMC.

4.3.3.2 SD Model When Applying Add-on Pattern

Next, Fig. 4.16 shows the SD model when Add-on pattern is applied to the business model with Self-service pattern and redelivery is charged separately.

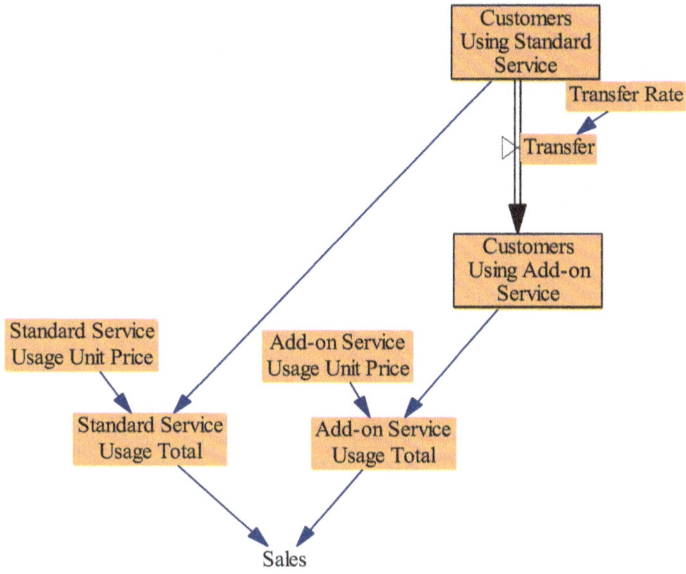

Fig. 4.13 SD model component of Add-on pattern (shaded in orange). To illustrate this with the example of a home delivery service, if the redelivery service when the customer is not at home, which has been provided under the standard charge, is made a separate charge, the customer first pays the standard charge (standard service usage unit price) when he or she applies for home delivery, and then pays a separate charge (add-on service usage unit price) for redelivery on delivery when redelivery becomes necessary) for redelivery with cash on delivery. The SD model shows that the total amount of the use of the standard service and the total amount of the use of the add-on service can be calculated according to the percentage of the total number of customers who use the delivery service (standard service customers) who use the redelivery service (add-on service customers). This SD model shows that the sum of the two is the sales

Newly added elements to the SD model when Self-service pattern is applied are shown in red. In addition, the elements where the SD model component of Add-on pattern is applied are shaded in orange. The cause–effect relations specific to Add-on pattern are also included in this diagram, and they are also indicated by the orange relation lines based on the results of identification by CER on BMC. The interactions between Self-service pattern and Add-on pattern identified in the interaction diagram are also represented by the red relation lines in the SD model one by one.

4.3.4 *Reflecting the Interaction between BMPs*

In Sect. 4.2.3, we discussed an example of identifying the interaction between Self-service pattern and Add-on pattern in the delivery industry. Essentially, Self-service pattern and Add-on pattern have their own, internal, cause–effect relations between

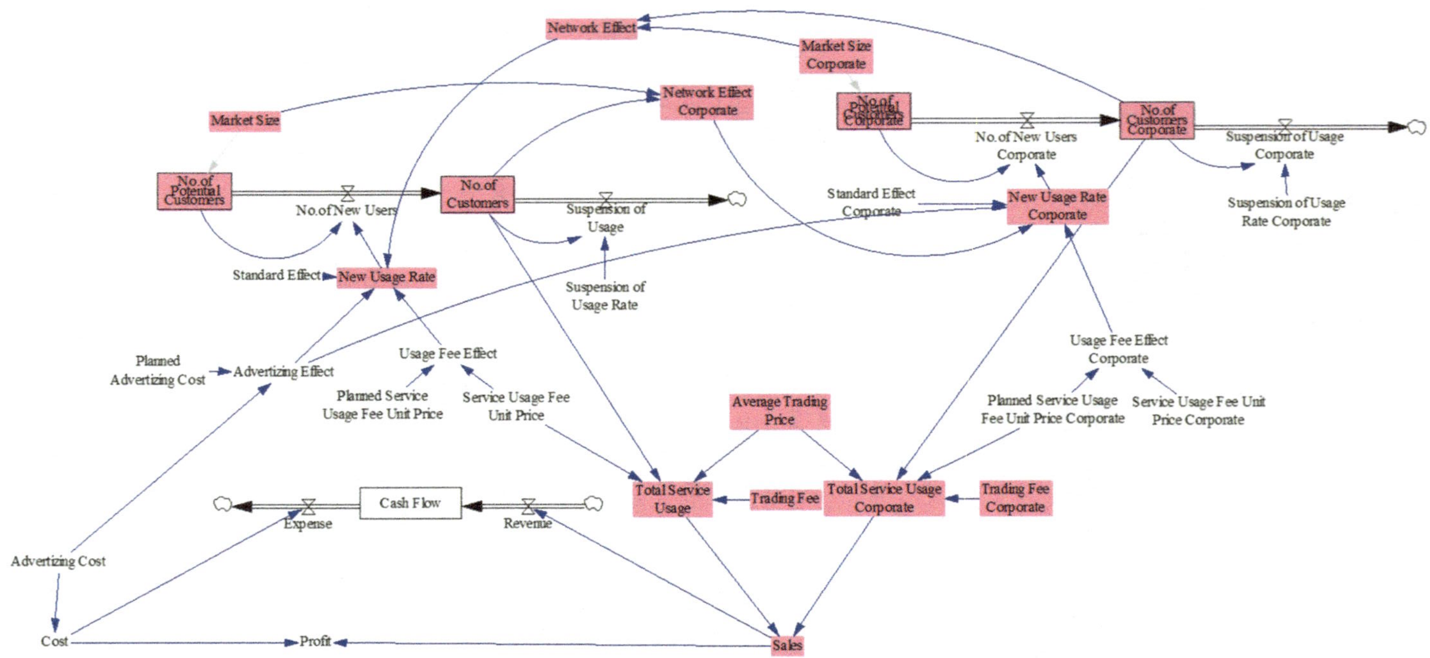

Fig. 4.14 SD model framework for Two-sided Market pattern (shaded in pink). For an e-commerce site with a membership system, if the number of stores in the site is large, the site will be attractive to customers (network effect), and the number of customers visiting the site will increase (new usage rate), while if the number of customers visiting the site increases, the site will be attractive to stores (network effect_companies), and more stores will open stores (new usage rate_companies). On the other hand, if the number of customers visiting the site increases, the site will become more attractive to stores (network effect_companies) and more stores will open (new usage rate_companies). This SD model shows that the operator of an e-commerce site earns sales by receiving admission fees and commissions based on the transaction amount from both customers and stores

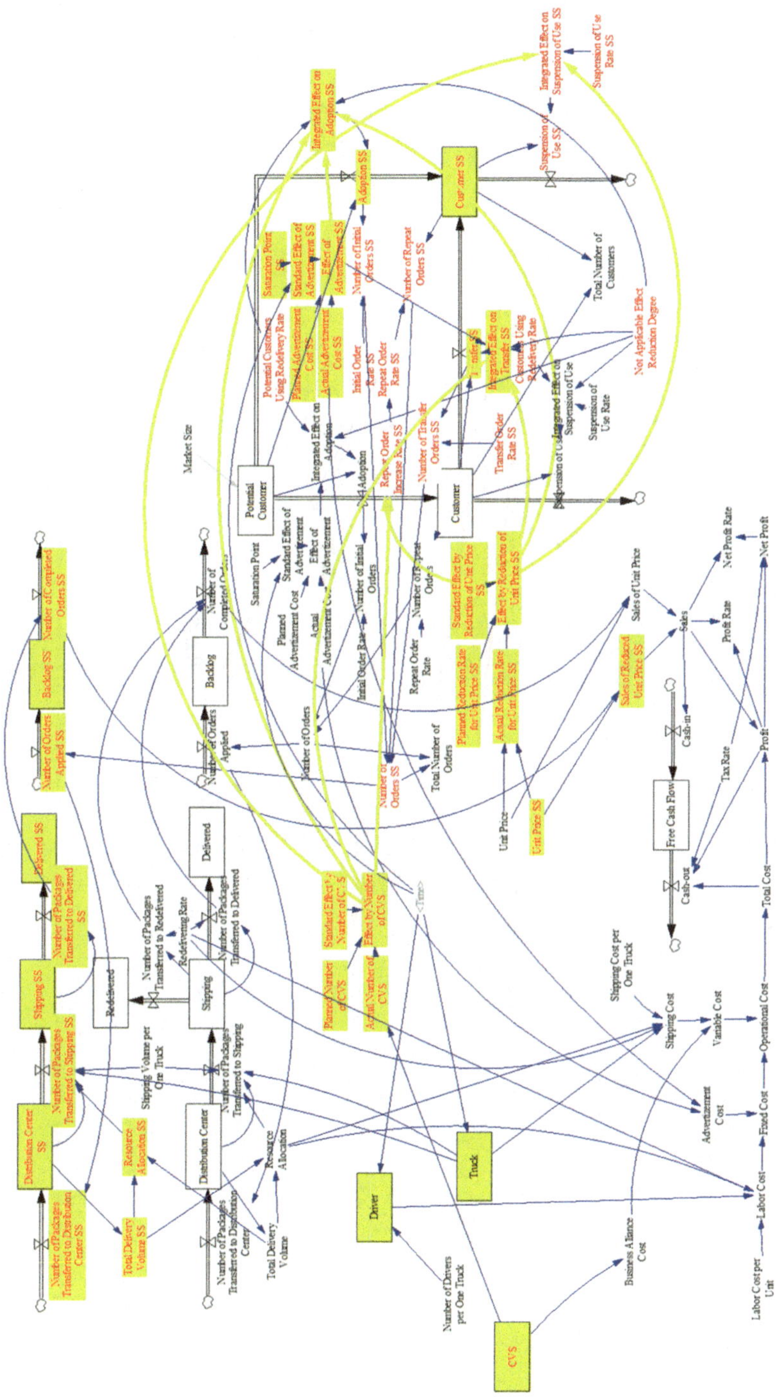

Fig. 4.15 SD model for the home delivery industry: after applying Self-service pattern

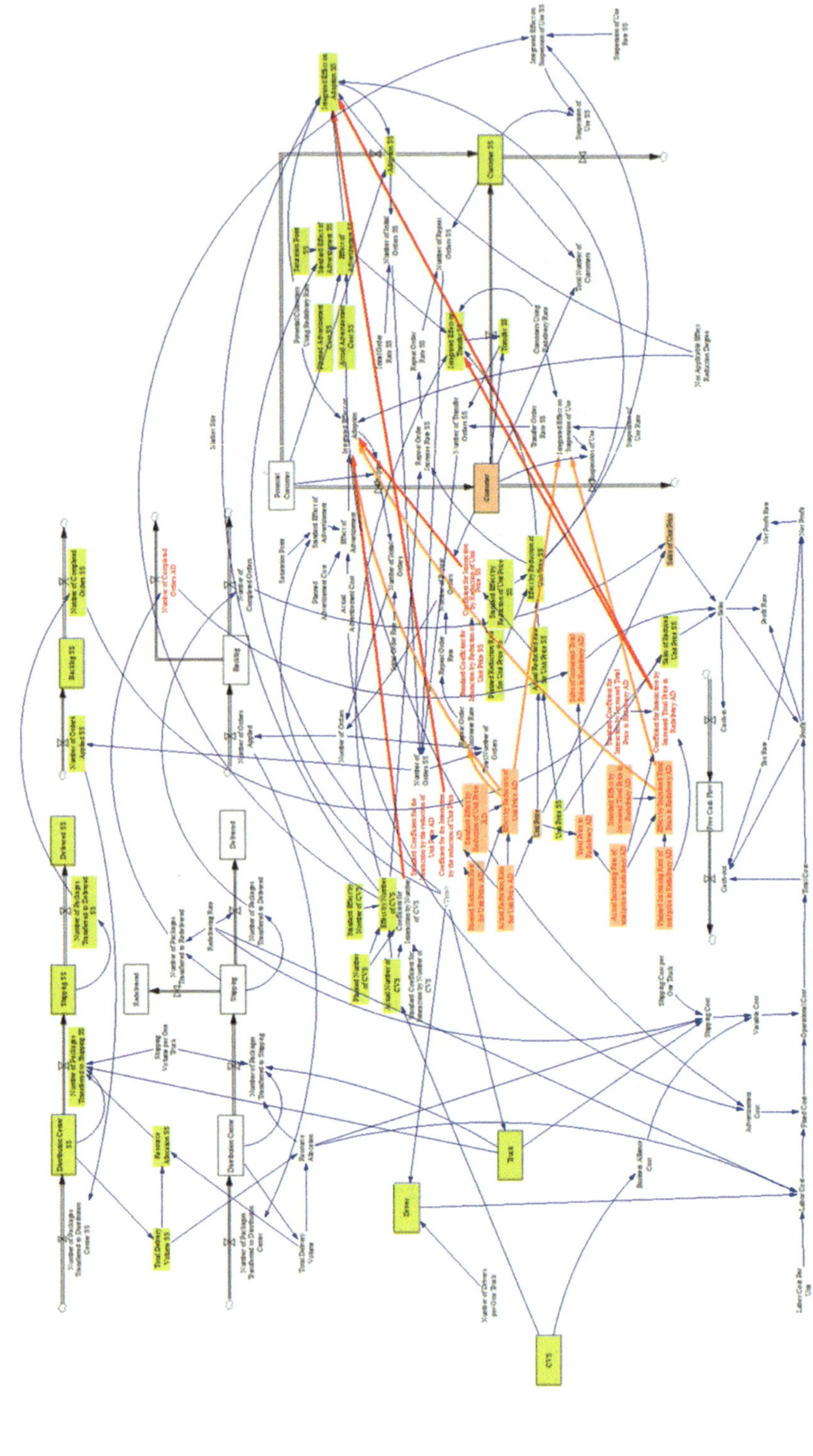

Fig. 4.16 SD model for the home delivery industry: after applying Add-on pattern

the measures and their result states. Therefore, when two BMPs are applied, two kinds of forces, the cause–effect relations inherent in the BMPs and the interaction between the BMPs, work together between the measures and their result states, and appear as an integrated effect. In order to correctly reflect the interaction between BMPs in the SD model, it is necessary to formalize the model with the integrated effect in mind. Based on the fact that the interaction acts on flows in Fig. 4.6, this section discusses the case where the interaction acts on existing flows in the SD model and the case where a new flow is created between stocks, and explains the rules for the formulation of the former case in Sect. 4.3.4.1 and the latter case in Sect. 4.3.4.2. The latter case is explained in Sect. 4.3.4.2.

Here, instead of re-creating the CER on BMC for the case where two BMPs are applied and constructing the SD model from it, we adopt the method of reflecting the results of analysis through the creation of the CER on BMC and interaction diagram for the case where the second BMP is applied to the SD model constructed from the application of the first BMP. There are two reasons for adopting this method.

1. Since the application of BMPs requires the implementation of measures, which entails huge costs, even if multiple BMPs are listed as candidates at the stage of consideration, the actual application of BMPs will involve prioritizing BMPs, applying BMPs one by one, confirming their effectiveness, and adjusting the allocation of resources. In such a situation, if the SD model is constructed by re-creating the CER on BMC, the SD model constructed when the first BMP is applied and the SD model constructed when the two BMPs are applied together will not be continuous. When simulating the gradual transition to a new business model as described above, it becomes difficult to execute their simulation.
2. In the method presented in this chapter, the results of the analysis are reflected in the original SD model, so that a single SD model is evolved, which, contrary to 1), facilitates the continuous execution of the simulation. In addition, from the results of the analysis by CER on BMC and interaction diagrams, the cause–effect relations and interactions identified in the analysis are reflected in the SD model on an one-to-one basis. The original purpose of the simulation, i.e., the formulation of a business plan based on the prediction, can be facilitated as a whole.

4.3.4.1 When an Interaction Acts on an Existing Flow

In the SD model, it is considered the effect of one element X on another element Y based on the cause–effect relation as giving the expected effect on Y (which is called the standard effect) when X takes a predetermined planning value. Specifically, as an example, the following equation is used to formalize the effect $\varepsilon_{X\to Y}$ of an element X on another element Y (the following equation is for the case where positive causality exists).

$$\varepsilon_{X\to Y} = \alpha_{X\to Y} \times \left\{1 - \exp\left(-X_{\text{actual value}}/X_{\text{planned value}}\right)\right\} / \{1 - \exp(-1)\}$$

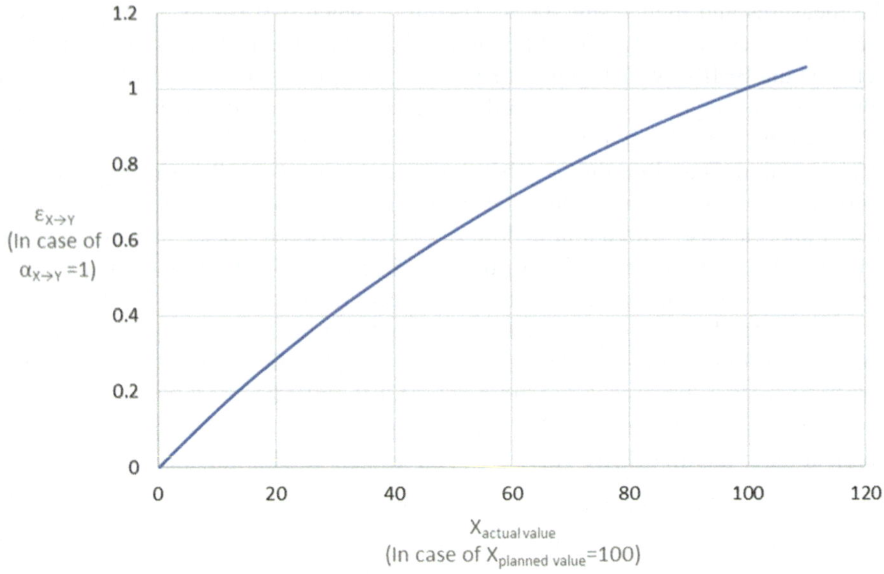

Fig. 4.17 Shape of the graph of $\varepsilon_{X \to Y}$

$X_{\text{actual value}}$: actual value of element X (positive real number).
$X_{\text{planned value}}$: planned value of element X (positive real number).
$\alpha_{X \to Y}$: Standard effect of element X on element Y.

If we set an appropriate value for the $X_{\text{planned value}}$ and take the $X_{\text{actual value}}$ as the horizontal axis and $\varepsilon_{X \to Y}$ as as the vertical axis, the shape of the graph becomes a monotonically increasing curve as shown in Fig. 4.17.

Now, suppose that X and Y are elements of a pattern Pa, and that X has an effect on Y according to the cause–effect relations in Pa. In addition to X, there may be a situation in which an element Z has an effect on Y based on another cause–effect relation. We denote the effect of Z by $\varepsilon_{Z \to Y}$ and the integrated effect of X and Z on Y by $\varepsilon_{X+Z \to Y}$. In order to formalize this integrated effect, we assume that there are three types of effects: main effects and promotion/suppression effects. The meaning of each effect is as follows.

Main effect: A quantitative expression of the effect.
Promotive effect: Based on the main effect, the degree to which the effect is promoted is expressed as a rate (a number greater than 1).
Suppressive effect: Based on the main effect, the degree to which the effect is suppressed is expressed as a rate (the number is smaller than 1).

Let $\varepsilon_{X \to Y}$ be the main effect and $\varepsilon_{Z \to Y}$ be the promotive/suppressive effect, then the integrated effect $\varepsilon_{X+Z \to Y}$ can be formalized as follows

$$\varepsilon_{X+Z \to Y} = \varepsilon_{X \to Y} \times \varepsilon_{Z \to Y}$$

There may be more than one element that gives the promotive/suppressive effect, in which case the effectε is multiplied by the number of such elements.

Nextly, when the pattern Pb is also applied, we assume that, in addition to X, the element V of Pb also interacts with Y (which is a flow). In this chapter, this interaction is denoted by the interaction coefficient $\iota_{V \to Y}$ from the element V in Pb to the element Y in Pa, and is expressed as a rate of the degree to promote or suppress the effect to the underlying main effect. In this way, the interaction from element V in Pb to element Y in Pa is incorporated into the above integrated effect as follows

$$\varepsilon_{X+Z \to Y} \times \iota_{V \to Y} = \varepsilon_{X \to Y} \times \varepsilon_{Z \to Y} \times \iota_{V \to Y}$$

The values of the interaction coefficients are formalized by adopting the relationship between the planned values and the standard interaction coefficients as a rule, as in the case of cause–effect relations. The following is an example of the formulation.

$$\iota_{V \to Y} = \exp\left\{\left(V_{\text{actual value}} / V_{\text{planned value}}\right) \times \ln\left(\rho_{V \to Y}\right)\right\}$$

$V_{\text{actual value}}$: actual value of element V (positive real number).
$V_{\text{planned value}}$: planned value of element V (positive real number).
$\rho_{V \to Y}$: Standard interaction coefficient of element V of pattern Pb on element Y of pattern Pa.

If we set an appropriate value for the $V_{\text{planned value}}$ and take the $V_{\text{actual value}}$ as the horizontal axis and $\iota_{V \to Y}$ as the vertical axis, the shape of the graph becomes an exponential curve with $\iota_{V \to Y} = 1$ when the $V_{\text{actual value}} = 0$ and $\iota_{V \to Y} = \rho_{V \to Y}$ when the $V_{\text{actual value}} = V_{\text{planned value}}$, as shown in Fig. 4.18. If $\rho_{V \to Y} > 1$, then $\iota_{V \to Y} > 1$ (promotive effect), and if $\rho_{V \to Y} < 1$, then $\iota_{V \to Y} < 1$ (suppressive effect).

When element U in pattern Pb also interacts with Y, the integration effect $\iota_{U+V \to Y}$ by elements U and V is formalized as follows.

$$\iota_{U+V \to Y} = \iota_{V \to Y} \times \iota_{U \to Y}$$

Therefore, the integrated effect, together with the cause–effect relation, is as follows.

$$\varepsilon_{X+Z \to Y} \times \iota_{U+V \to Y} = \varepsilon_{X \to Y} \times \varepsilon_{Z \to Y} \times \iota_{V \to Y} \times \iota_{U \to Y}$$

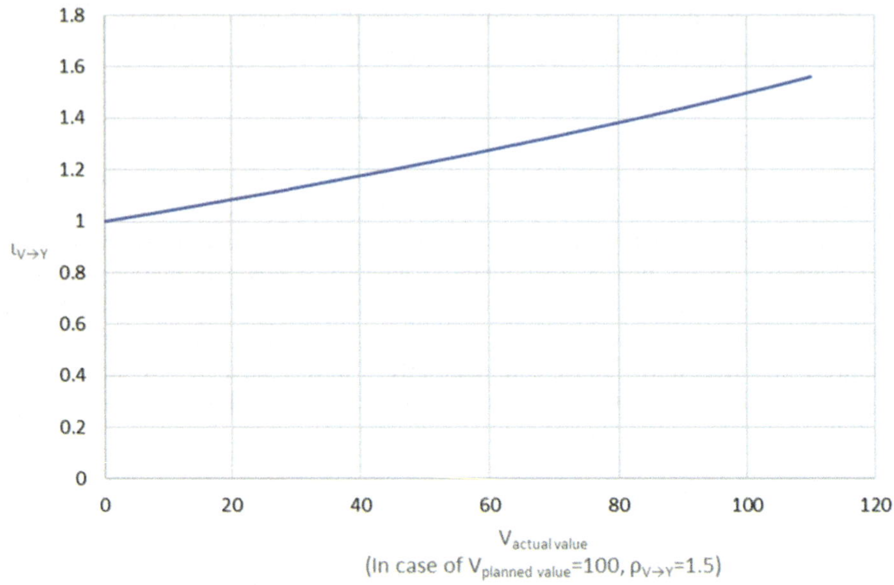

Fig. 4.18 Shape of the graph of the interaction coefficient $\iota_{V \to Y}$

Now, the integrated effect of all the elements in the pattern Pa that effect the element Y is as follows.

$$\varepsilon_{\mathrm{Pa} \to Y}$$

On the other hand, the integration effect on element Y by all elements in the pattern Pb that interact with element Y is as follows.

$$\iota_{\mathrm{Pb} \to Y}$$

We also denote the integration effect on Y by all elements in the pattern Pc that interact with the element Y is as follows.

$$\iota_{\mathrm{Pc} \to Y}$$

From the above, the integration effect of pattern Pa, pattern Pb, and pattern Pc on Y is as follows.

$$\varepsilon_{\mathrm{Pa} \to Y} \times \iota_{\mathrm{Pb} \to Y} \times \iota_{\mathrm{Pc} \to Y}$$

Also, when there are three or more BMPs that interact with each other, if any one of them is denoted by $p*$, the integration effect is expressed by multiplying the interaction coefficient $\iota_{P* \to Y}$ in the same way as above. Since the commutative and associative laws are valid in multiplication, the assumption of this formulation is that the interaction diagram is created between the pattern Pa and any pattern $P*$ that interacts with the elements of Pa, then the same integration effect is eventually obtained by multiplying the results of the analysis sequentially in the equation of the integration effect, regardless of the order in which the BMPs are introduced.

However, from the viewpoint of real business scene, integration applying a large number of BMPs that interact with a certain variable based on the policy formulated for the opportunity or threat (here, we consider the situation where the interaction is recognized and intentionally combined as a measure) is inefficient and cannot be recommended as a new business model, because it increases the number of measures and requires a large amount of cost for change and operation, even if the goal for the variable can be achieved. If the goal cannot be achieved by applying one BMP, the content of the measures should be improved within the BMP, rather than applying a new BMP.

4.3.4.2 When an Interaction Creates a New Flow between Stocks

Interactions between BMPs may create new flows between stocks where no flows exist originally. Unlike the transition based on the cause–effect relations inherent in BMPs, such flow-based transition requires a different formulation from the one in Sect. 4.3.4.1, because there is no flow between the stocks and the effect cannot be formalized yet.

Again, let X and Y be elements of the pattern Pa, and let X have an effect on Y based on the cause–effect relations in Pa. Let V and W be elements of the pattern Pb, and let V have an effect on W based on the cause–effect relations in Pb. In this case, the interaction of V causing the transition from Y to W is expressed by the coefficient $\iota_{V \to (Y \to W)}$, and the interaction of X causing the transition from W to Y is expressed by the coefficient $\iota_{X \to (W \to Y)}$. The following is an example of the formulation.

$$\iota_{V \to (Y \to W)} = 1 - \exp\left(V_{\text{actual value}} \times \log\left(_{1-\rho V \to (Y \to W)}\right) / V_{\text{planned value}} \right)$$

$V_{\text{actual value}}$: actual value of element V (positive real number)
$V_{\text{planned value}}$: planned value of element V (positive real number)
$\rho_{V \to (Y \to W)}$: standard interaction coefficient causing a transition from element Y of pattern Pa to element W of pattern Pb by element V of pattern Pb.

This means that if we set an appropriate value for the $V_{\text{planned value}}$ and take the $V_{\text{actual value}}$ as the horizontal axis and $\iota_{V \to (Y \to W)}$ as the vertical axis, the shape of the graph will be an asymptote starting from $\iota_{V \to (Y \to W)} = 0$ when the $V_{\text{actual value}} = 0$, $\iota_{V \to (Y \to W)} = \rho_{V \to (Y \to W)}$ when the $V_{\text{actual value}} = V_{\text{planned value}}$, and $\iota_{V \to (Y \to W)} = 1$ when the $v_{\text{actual value}} \to \infty$, as shown in Fig. 4.19.

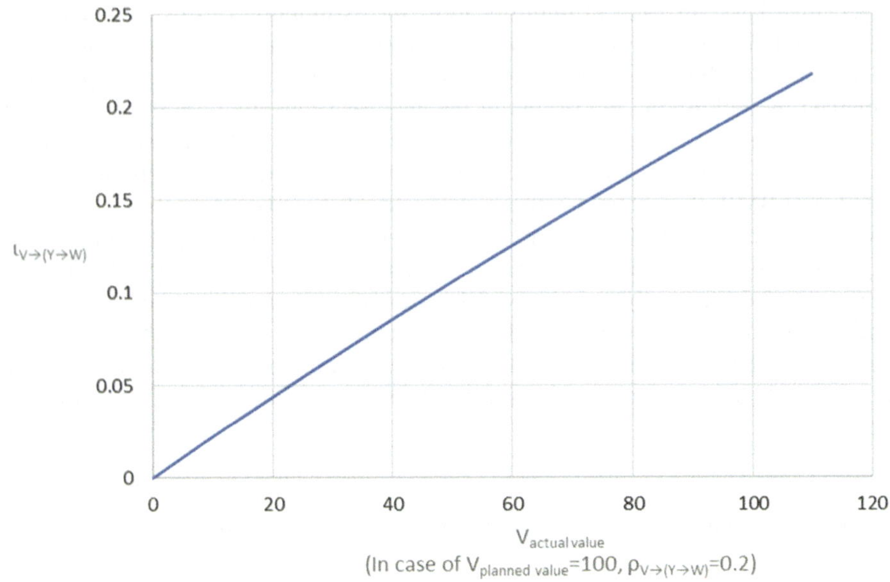

Fig. 4.19 Shape of the graph of the interaction coefficient $\iota_{V\to(Y\Rightarrow W)}$

This $\iota_{V \to (Y \to W)}$ means, as its main effect, the $Y_{\text{present value}} \times \iota_{V \to (Y \to W)}$ gives the specific number of transitions.

In this formulation, pattern Pa and pattern Pb are "taking" what is accumulated in *Y* through the action of *V*. However, there is another type of relation, such as "giving" relation that can be considered. In this chapter, however, the "taking" relation is formalized and explained, since it is the "taking" relation caused by the competition between BMPs that often occurs in practice.

Now, when the SD model shown in Fig. 4.16 is opened in Vensim, the dialog shown in Fig. 4.20 is opened for each stock, each flow, and each variable in the model through the mouse click operation. The equations for the effects and interaction coefficients shown in this section are set in the Equations column in the dialog, using the variables corresponding to the effects and interaction coefficients.

4.3.5 Running the Simulation

For the SD model built in the previous section, we run simulations using Vensim (Vensim, 2021). Then, the feasibility, profitability, and growth potential of the new business model are verified by visualizing and expressing in graphs the transition

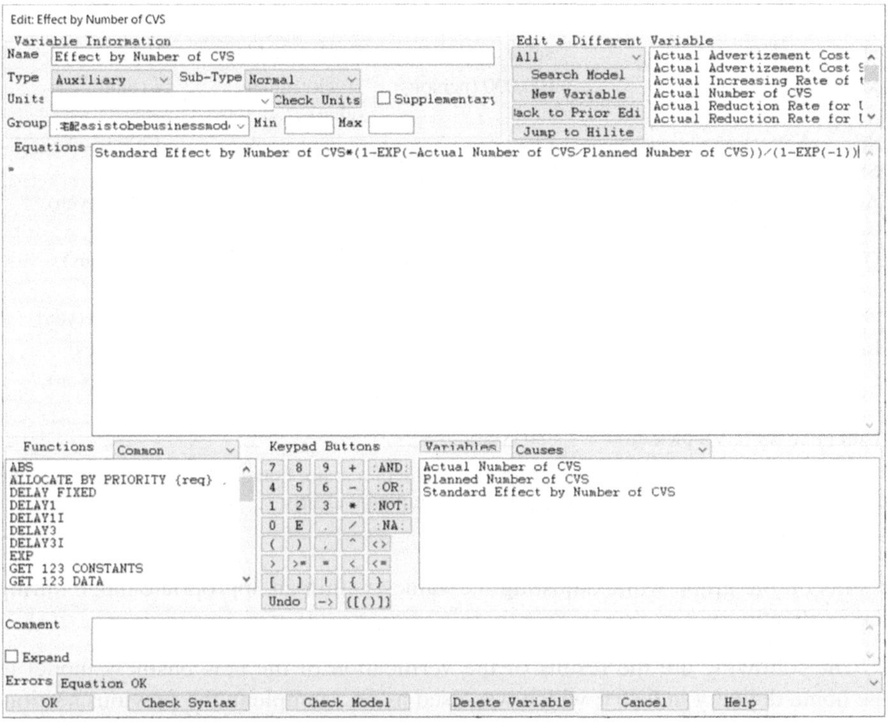

Fig. 4.20 The dialog where the "CVS number effect" is configured

of various indices that are incorporated as stocks and variables in the SD model. Specifically, the following points are used as criteria for confirmation.

Feasibility
There should be no sudden changes in the shape of the graphs of the amount of parts ordered, the amount of products produced, the amount of products delivered, etc., after a certain point in time, and these should be linked in the order of time so that there is no retention.

Profitability
Based on sales and expenses, profits, net profit margins, and cash flows should increase compared to the current business model, and there should be no sharp decline in these trends after a certain point.

Growth Potential
Sales, profits, and the number of customers adopting the new product/service are increasing steadily, and employment and capital assets are increasing to provide the new product/service.

However, since the tendency of these points may change significantly by changing the values of the stock and variables, it is necessary to repeatedly check the

Table 4.2 Values of stocks and variables defined in the SD model[a]

Stock variables	Value	Stock variables	Value
Market size	100,000 (persons)	Truck (initial value)	4 (units)
Standard effect of advertisement SS (CVS pick up)	1 (%)	Shipping volume per one truck	1200 (packages)
Actual advertisement cost SS (CVS pick up)	100,000 (yen)	Shipping cost per one truck	50,000 (yen)
Order rate	10 (%)	Number of drivers per truck	2 (persons)
Order rate SS (CVS pick up)	15 (%)	Labor cost per unit	200,000 (yen)
Re-delivering rate	30 (%)	CVS	50 (stores)
Unit Price	930 (yen)	Business Alliance cost	50,000 (yen)
Unit Price SS (CVS pick up)	880 (yen)		

[a]The unit price of the fee is based on the fee schedule of a major home delivery company, and the other values are set as examples to show the SD simulation clearly

results on the graph while adjusting the values within an appropriate range during the verification.

The following are the results of the verification of the new business model in the home delivery industry, which was used as an example in the previous section, based on the above criteria. In this section, the values of stocks and variables are set as shown in Table 4.2.

First, the results of the SD simulation of the current business model are shown in Fig. 4.21 for the change in sales over time.

Next, we show the results of the simulation of the SD model (Fig. 4.15) with Self-service pattern applied, with respect to the change in sales over time (Fig. 4.22). In this simulation, the SD model of the current business model (Fig. 4.12) was first run for 100 months, and then the number of customers was set as the initial value in the SD model applying Self-service pattern in the situation where customers were transited to each segment. In addition, except for the advertising effect, which is positioned as the main effect in the formulation of the effect, the simulation was run with the following settings: standard effect = 1.2 (for the promotive effect) or 0.8 (for the suppressive effect), and actual value = planned value.

As a result, it was confirmed that the launch of the package pickup service at CVS led to the transfer of customers from the existing service and the attraction of new customers who had not previously used the home delivery service, bringing about an increase in the total number of customers and a corresponding increase in sales. However, when the simulation was continued after 50 months, it was confirmed that the upward trend in sales was not permanent and, coupled with the change in the number of customers, the trend turned into a gradual decrease after 52 months.

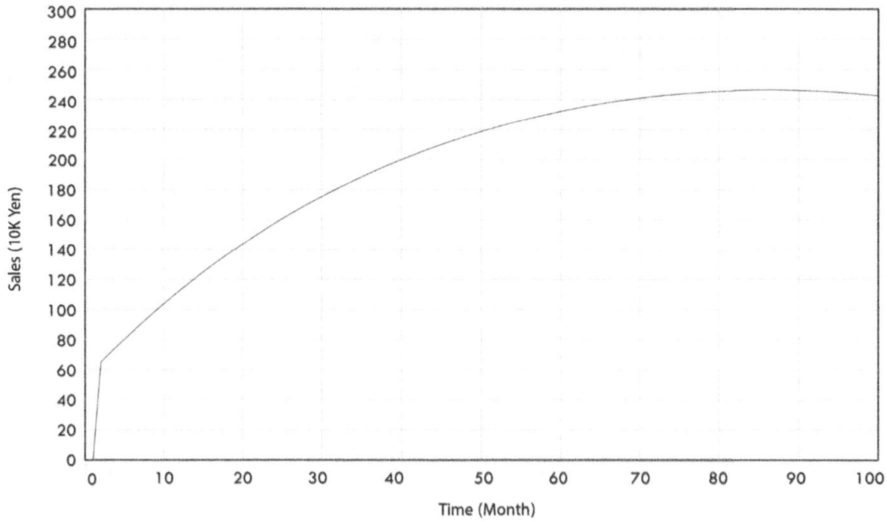

Fig. 4.21 Change in sales over time: current business model

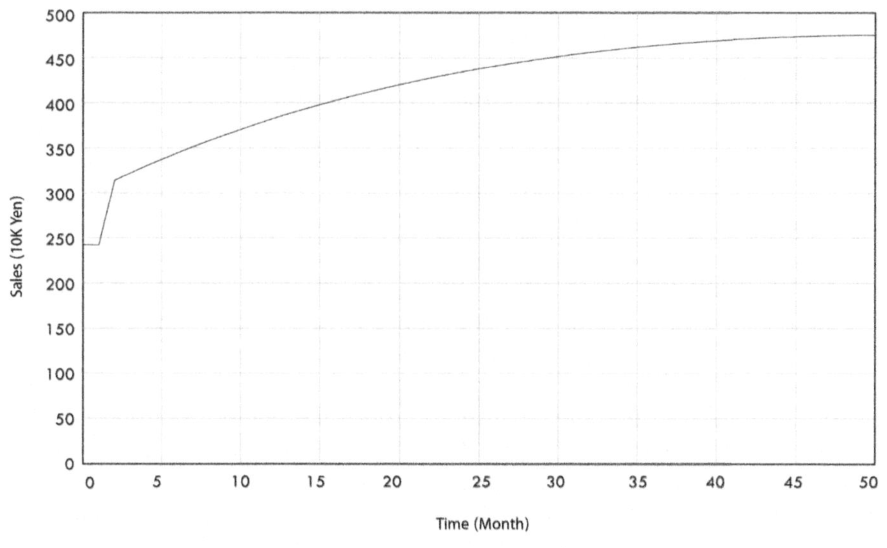

Fig. 4.22 Change in sales over time: after applying Self-service pattern

Furthermore, the results of the simulation of the SD model with Add-on pattern (Fig. 4.16) are shown in Fig. 4.23 for the change in sales over time. In this simulation, after 50 months of simulating the SD model with Self-service pattern (Fig. 4.15), we set the number of customers transited to each segment as the initial value in the SD model with Add-on pattern (Fig. 4.16), and simulated it for another

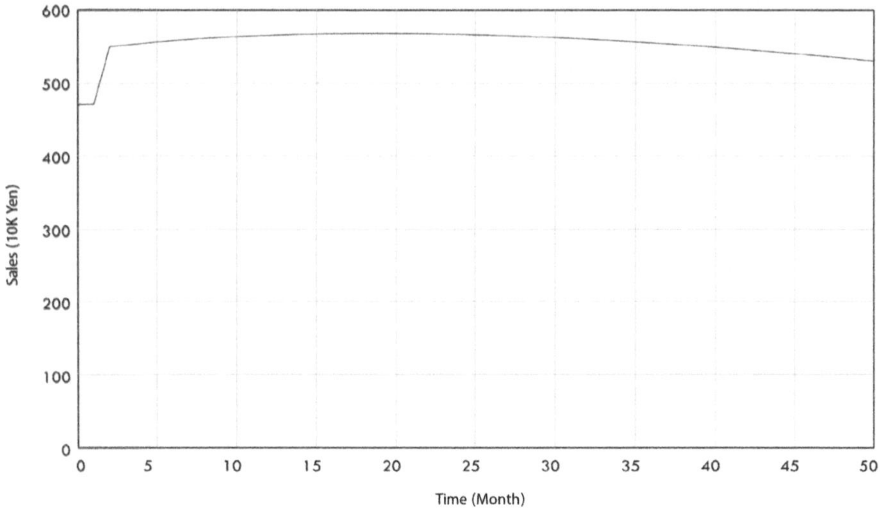

Fig. 4.23 Change in sales over time: after applying Add-on pattern

50 months to check the changes in sales and the number of customers. We confirmed the changes in sales and the number of customers. Again, in the formulation of the effects and interactions, except for the advertising effect, which was positioned as the main effect, the other effects were generally defined as follows: standard effect = 1.2 (for the promotive effect) or 0.8 (for the suppressive effect), standard interaction coefficient = 1.2 (for the promotive effect) or 1/1.2 (for the suppressive effect), and actual value = planned value.

As a result, it is confirmed that the sales can be further increased if the extra charge for redelivery is added at the same time, compared to the case where only the package pickup service at CVS is continued. This is due to the cause–effect relations within BMPs identified on the CER on BMC and the interactions between BMPs identified on the interaction diagram.

We also confirmed that the redelivery rate can be reduced from the business process of home delivery by growing the business model from the current model to the model applying Self-service pattern and then to the model applying Add-on pattern. The results are shown in Table 4.3. The number of delivered packages is shown on the left-hand side of the table when the current model is used to continue the delivery service after 200 months, and the number of delivered packages is shown in the center of the table when the model with only Self-service pattern introduced after 100 months is used to continue the delivery service after 200 months. The case in which the delivery service is continued until the end of 200 months with the model that also applies Add-on pattern introduced at the end of 150 months is shown on the right side of the table.

Table 4.3 Number of delivered packages accumulated in each stock at the end of 200 months

Current business model		New business model (self-service pattern only)		New business model (also add-on pattern)	
Stock name	Number of packages delivered	Stock name	Number of packages delivered	Stock name	Number of packages delivered
Delivered	319,363	Delivered	307,599	Delivered	318,301
Redelivered	136,870	Redelivered	131,829	Redelivered	136,415
–	–	Delivered SS	248,848	Delivered SS	294,015
Total amount	456,233	Total amount	688,276	Total amount	748,731

This shows that the total number of delivered packages increases by 44% when only Self-service pattern is applied, and by 71% when Add-on pattern is also applied, compared to the current model. In spite of this increase, the percentage of re-delivery is 19% when only Self-service pattern is applied, and only 18% when Add-on pattern is also applied.

In this way, we will record the values of stocks and variables defined in the SD model as well as the indicators to be watched in the new business model and the indicators that serve as criteria for feasibility, profitability, and growth, when the initial targets are achieved. For example, in the above verification of the business model of the home delivery industry, the number of trucks is adjusted appropriately on the time axis in response to changes in the volume of packages to prevent packages from stagnating and to avoid reducing profits. By recording the results of such adjustments, when the goal is achieved, it will be possible to show how it was achieved by combining the SD model and the values, which will be effective information as evidence of the conceptual design and its effects.

4.3.6 Running SD Simulation with Monte Carlo Simulation

The standard effect α and the standard interaction coefficient ρ defined in sect. 4.3.4 are difficult to be defined as fixed values in practice, and it is appropriate to treat them as random variables assuming a certain distribution when running the simulation. Therefore, we run a combination of SD simulation and Monte Carlo simulation (hereafter referred to as MC simulation), and output the graphs of the temporal changes of each variable with confidence intervals. The Monte Carlo sensitivity analysis provided by Vensim (Vensim, 2021) can be used as a tool to perform this simulation.

In such a case, we compare the sales and profits when the business is operated under the new business model applying BMP, with those when the business is operated under the current business model, and confirm that the sales and profits are higher than those under the current business model, including the variability expressed by the confidence interval, to determine whether or not the new business model should be introduced. Alternatively, based on the Central Limit Theorem, by increasing the number of times the MC simulation is run, the average value of the results will approach the expected value, so the expected value is calculated using the tool and compared with the current value to determine whether the new business model should be introduced or not.

In the simulations up to this point, we have shown the results of runs with fixed values of standard effects and standard interaction coefficients. However, since this is difficult to predict for a new business model without any previous experience, we will treat these variables as random variables assuming a certain distribution. In this chapter, we experimentally treated them as uniform distributions with a width of ±0.1 for the fixed values of 1.2 or 0.8. [ii)] The number of trials was set to 200, then simulations were run.

Here, Fig. 4.24 shows a graph outputting the change in sales over time with confidence intervals for the model applying only Self-service pattern in Fig. 4.15.

The same simulation was performed for the model applying Self-service and Add-on patterns in Fig. 4.16, and the graph outputting the change in sales over time is shown in Fig. 4.25.

As mentioned before, Time = 0 in the graph of Fig. 4.25 corresponds to the time of Time = 50 in the graph of Fig. 4.24. Based on this correspondence, when comparing the two graphs, the interval in Fig. 4.25 exceeds the interval in Fig. 4.24 even when viewed with a confidence interval of 100%, so we can conclude that a business model that allows both home delivery and CVS pickup and charges extra for redelivery should be introduced.

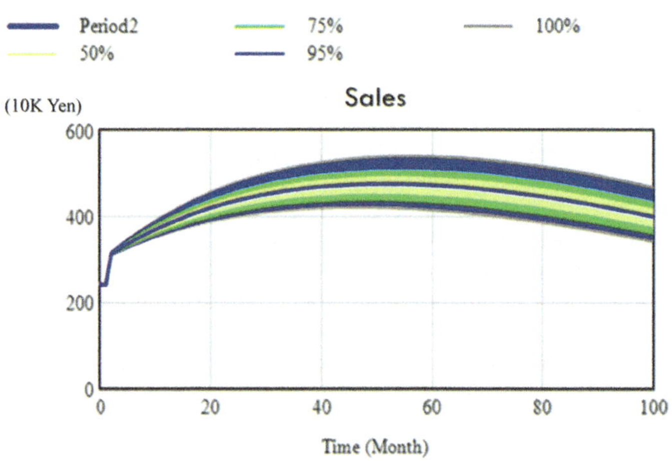

Fig. 4.24 Example of SD simulation running combined with MC simulation

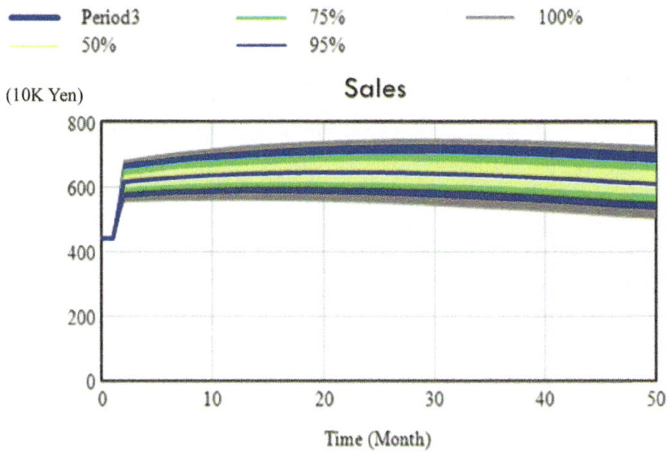

Fig. 4.25 Change in sales over time: Self-service and Add-on patterns

4.4 Future Issues

In this chapter, the process for conceptual design and verification is defined in Sections 4.2 and 4.3 for the creation of new business models. Based on the integration of BMC and SD (Minato, 2013) and the advocacy of SWOT on BMC and AF on BMC (Osterwalder & Pignol, 2012), the original point of this method is that the process is defined in a way that newly incorporates BMP.

In the conceptual design phase, through SWOT on BMC and AF on BMC, it is considered that what to focus on and what measures to take may differ from person to person, even if they are dealing with the same subject under the same circumstances. However, once the conceptual design has been established, in the subsequent verification phase, it is required that the final result should be the same regardless of who executes it, without any room for arbitrariness. In addition, we believe that it is important to assert the effectiveness of the process that it can be realized in practice by following the process proposed in this chapter.

In this section, we list the points that we considered to be problems in obtaining the same result all at once when running simulations with SD in the verification phase according to the proposed process, and we clarify the measures adopted in this process and the issues to be solved. We also summarize the issues that became apparent during the conceptual design and verification of a new business model by applying multiple BMPs according to the process proposed in this chapter. In addition, we discuss the issues to be addressed when expanding the range of BMPs to be applied beyond (Gassmann et al., 2016).

We considered the following four points to be issues.

1. Conversion from conceptual design results on BMC to SD model.
2. Representation of the interaction between BMPs in the SD model.

3. Introduction of blocks and layers for building SD models and providing views for them.
4. Unify templates and extend processes to other BMPs.

In the following sections, we describe the measures against them and issues to be remained.

4.4.1 Conversion from Conceptual Design Results on BMC to SD Model

In our process, as explained in Sect. 4.3.2, we standardized how to make changes to the current SD model according to the types of measures ("remove," "reduce," "increase," and "add") described separately in the BMC.

In addition, if there are parts of BMP that can be formalized as SD models, we have made them into parts as SD model templates/components/frameworks.

In this way, when building an SD model corresponding to a new business model from the results of conceptual design on the BMC, the changes to be made to the current SD model are standardized, so that the SD model built does not differ from person to person.

By the way, although we mentioned that the SD model is subdivided for each BMP, to be precise, there are some BMPs that can not be made into parts. For example, Open Business pattern and Orchestrator pattern, etc., define business architecture from their descriptions, and even if a concrete scenario is clarified, the SD model built based on it remains an instance for each case, and cannot be reused as an SD model. Therefore, if we were to apply these BMPs, we would not be able to standardize the changes to be made to the SD model.

Whether or not the 55 BMPs are subject to componentization, and if so, whether or not only the aforementioned three types of BMPs are all of it, and which BMPs belong to which types, are issues that require further investigation. For BMPs that can be made into parts, we will continue our efforts to describe their composition and accumulate them as actual parts, as shown in Figs. 4.12, 4.13, and 4.14.

4.4.2 Representation of the Interaction between BMPs in the SD Model

When the interaction between BMPs acts on the flow from one stock to another, it does not act on all the "people," "things," and so on, stored in the starting stock in general. Instead, the action may be applied only to those that satisfy specific conditions. For example, using the example of the interaction diagram in Fig. 4.6, if redelivery is charged separately, the number of potential customers who need redelivery will increase in transition to CVS pickup customers, but on the other hand, the number of customers who do not need redelivery will not increase, because the price per unit will be reduced by introducing the extra charge for redelivery. Therefore, in Fig. 4.6, the "potential customers" are subdivided into "redelivery: required" and "redelivery: not required," and the interaction between them is shown. The interaction is expressed in the form that the action is applied only to "redelivery: required."

However, even if this relationship is visualized and expressed in the interaction diagram, it cannot be reflected in the diagram of the SD model when it is built in the verification phase. Instead, it is all replaced by an expression in the stock and set in the Equations column in the dialog as shown in Fig. 4.20, and are hidden inside the SD model. Then, it will be difficult for a third party to understand the interactions between BMPs when he or she sees only the diagram of the SD model later.

By the way, when multiple BMPs are applied in combination, the BMPs are considered to have a relationship that produces a synergistic effect (i.e., a relationship where $\alpha > 1$ and $\iota > 1$, or $\alpha < 1$ and $\iota < 1$). Figure 4.6 also shows that the application of Self-service pattern allows the customer to pick up the package at CVS at any time, and reduces the unit cost of the CVS pickup fee, so that, for example, a "potential customers" ("redelivery: required") will increase in transition to "CVS customers," moreover, the application of Add-on pattern will increase the total amount of charges for redelivery, which will further promote the migration from "potential customers" ("redelivery: required") to "CVS customers."

Therefore, if we can represent the general relationships among BMPs at a meta-level in the upper hierarchy of the SD model, and at the same time, if we can represent the target of action in the stock by subdividing it according to the conditions. If the simulation can be performed based on these contents, the results of the conceptual design with the BMP can be clearly reflected in the SD model and the verification can be performed efficiently.

4.4.3 Introduction of Blocks and Layers for Building SD Models and Providing Views

In the current SD model, all the elements are represented flatly on a single canvas, and as the number of elements increases, as shown in Fig. 4.16, the lines of relationship become jumbled and difficult to read, making it difficult to understand the details of the model from a single diagram.

In this research, BMC is introduced at the conceptual design phase, and BMP is used as a part at the SD model building phase. Therefore, if the concept of blocks defined by BMC and the concept of layers for building SD models through step-by-step detailing of abstract elements constituting the part are introduced in the tool and they can be focused on by switching views, the thinking process of building SD models and the work on the tool will be seamlessly connected, and efficiency will be improved.

For example, the following functions can be considered.

- You can switch from a view that displays the entire SD model to a view that divides the elements of the SD model into nine blocks of the BMC and displays the elements that belong to each block.
- You can switch from a view that shows the entire SD model to a view that shows the elements for each BMP applied to it.
- In the BMP, by switching the view, you can move to each layer that records the process of step-by-step detailing of abstract elements.
- In order to be able to connect relationship lines between different blocks or different BMPs, the tool will be displayed in two screens, and the selected two blocks or two BMPs will be displayed respectively in each of them, and the relationship lines can be connected between them.

By the way, the above functions are based on the perspective of supporting SD models building, but if the tools are equipped with these functions, we believe that they will also be useful for third parties to understand the contents of SD models even after building has been finished.

4.4.4 Unify Templates and Extend Processes to Other BMPs

In this chapter to ensure the generality of the proposed process by making it applicable to as many business fields as possible, we adopt a pattern of generic common elements that are not specific to a particular industry or social initiative (Gassmann et al., 2016) in this study. However, efforts to extract patterns in business models are not limited to (Gassmann et al., 2016; Remane et al., 2017), and we will continue to study processes and patterns so that other patterns can be applied. Specifically, in order to expand the scope of patterns, we need to address the following points.

1. (Gassmann et al., 2016) is characterized by the fact that the composition of each pattern is described according to a unified template. In order to guarantee the uniqueness of the result of a process, the pattern should ideally be able to be read and judged from its description in a standardized manner. However, for some patterns other than those in (Gassmann et al., 2016), it is difficult to determine the applicability of a pattern based on its description alone, and in such cases, it is necessary to redescribe the pattern according to a uniform template, as in (Gassmann et al., 2016).
2. Some of the patterns are specific to particular industries or social initiatives, and while these patterns have a narrower scope of application, they tend to be deeply embedded in the industries or initiatives. Therefore, we believe that the process will be applied in a different procedure and in a different context from the process proposed in this chapter, then the process itself needs to be extended or recreated for this purpose.

4.5 Related Research

In this section, we summarize the results of domestic and international research related to the research themes described in this chapter, while clarifying the differences between them.

In (Fritscher & Pigtraneur, 2014), Computer-aided business model design (CABMD) is proposed and the authors' ideas are summarized in the following three themes in order to make it accessible to people without special training.

1. What concepts are needed when dealing with changes from one business model to another?
2. How should the gradual visualization of business model evolution be done?
3. What design principles does CABMD need to meet to support the visualization and manipulation of business model evolution?

This study is the result of a basic investigation for a tool to realize CABMD. However, although BMC is adopted as a visual modeling method, BMP and simulation are not treated, and in particular, verification of feasibility by simulation is not considered.

In order to materialize the concept of such a tool, research and development is being conducted by (Kizuna & Okada, 2019; Kizuna et al., 2019; Kizuna et al., 2020) under the name of "Business Model CAD System." The business model CAD system has newly added original mechanisms, such as SD simulation and background mechanism library, and is equipped with verification functions such as feasibility. The background mechanism library embodies the cause–effect relations for the occurrence of side effects associated with the implementation of measures as SD model components in advance, making it possible to incorporate side effects that the business model design members are not aware of into the SD simulation. However, the issue is how much the background mechanism can be clarified and incorporated into the tool in advance, and there is a concern that in practice, it may be necessary to consider each situation according to the application. (Kimata & Okada, 2019) introduces a method for building a background mechanism library, which can be used as a reference. However, it is different from the approach of this chapter, which defines the interaction (or side effect) among BMPs as a task in the process, in the conceptual design phase, independently of the SD model, with BMPs as the unit of componentization.

The integration of BMC and SD simulation has been tackled since (Minato, 2013; Romero & Villalobos, 2015), before the "Business Model CAD System". Minato (2013) is the basis of this chapter, and we do not need to explain it again here. On the other hand, in (Romero & Villalobos, 2015), in order to convert the description in BMC into SD model, a metamodel of the components of BMC is created, and the components of SD model are also added as elements of the metamodel after being classified into their own types, and the correspondence between them is clarified. Then, the correspondence between them is clarified, and the approach is taken to mechanically transform the components of each block of the BMC. This approach is different from the conventional approach of building SD models based on cause–effect relations within the target (this chapter follows the conventional approach).

(Tauscher & Chafac, 2016) shows that by assuming various scenarios and running SD simulations for them, the performance of the company changes with each scenario selected. The point of pre-verification by simulation before moving to a new business model has the same way of thinking with the aim of this chapter, and the point of identifying various alternatives as scenarios is to be incorporated into the conceptual design phase of the process in this chapter.

By the way, this chapter focuses on the 55 BMPs in (Gassmann et al., 2016), but the most comprehensive and systematic summary of BMPs is in (Remane et al., 2017). It contains 182 BMPs, including the BMPs in (Gassmann et al., 2016). Thus, all BMPs in (Remane et al., 2017) have their original references, which are cited in it with only a brief description. However, even in the literature cited, there are some BMPs that do not contain enough information, and in order to be able to make application decisions for these 182 BMPs, we believe that it is necessary to expand the descriptions of BMPs according to a unified template, as described in Sect. 4.4.4.

A hierarchical classification of BMPs has also been presented in Weking et al. (2020). In this paper, the concept of classification has not been incorporated into the process, but if the number of BMPs to be studied increases, such a classification can be expected to be useful in making application decisions.

The integration of BMC and simulation is not limited to SD simulation, but has also been done in agent-based simulation (hereafter referred to as AB simulation), for example, in (Lieder et al., 2017), a situation where a business model is shifted from a linear economy to a circular economy is examined by AB simulation. We would like to work on the introduction of such different techniques so that we can use different simulation techniques according to the different characteristics of the target field.

In addition, an approach to support the analysis and design of business models with an extended version of the business model canvas and the application of artificial intelligence using prolog is proposed in (Hatakama, 2018), which we would like to refer to as a new development from this process.

4.6 Summary

In this chapter, we proposed a process to conceptually design a new business model at the conceptual stage and to verify its feasibility, profitability, and growth potential. In this process, BMC is adopted as the architecture of the business model, and based on the results of SWOT analysis of the current business model, a policy for the creation of a new business model is formulated by applying BMP. Then, for each block of the BMC, we write down the measures to be implemented in accordance with the AF. From this, the SD model is constructed and simulations are performed to verify the feasibility, profitability, and growth potential of the new business model.

The proposal in this chapter is based on the business model design evaluation method based on the integration of BMC and SD proposed by Minato (2013), and adds the following new elements.

1. A series of tasks from creation to verification of a new business model is defined as a process.
2. Incorporating BMP as a template for new business models and formulating policies.
3. For each block of the BMC, the method of building SD models is standardized.
4. For each block of the BMC, the method of building the extended part of the SD model is standardized according to the results of AF application.
5. For each BMP, parts that can be formalized as SD model are made and applied to extend the SD model.
6. When multiple BMPs are applied, the tasks to be performed in conceptual design and the rules to be followed in verification are shown.

Finally, in order to obtain the same result uniquely in the verification phase, we listed the points that we considered to be problems, the measures adopted in this process to solve them, and the issues to be remained. In addition, we summarized the issues that became apparent during the conceptual design and verification of new business models by applying multiple BMPs according to the process proposed in this chapter. In addition, we discuss the issues to be addressed when expanding the range of BMPs to be applied.

We considered the following four points to be issues.

1. Conversion from conceptual design results on BMC to SD model.
2. Representation of the interaction between BMPs in the SD model.
3. Introduction of blocks and layers for building SD models, and providing views for them.
4. Unify templates and extend processes to other BMPs.

References

Amazon hub. (2021). Retrieved September 13, 2021, from https://www.amazon.co.jp/b?ie=UTF8&node=7083387051

Fritscher, B., & Pigtraneur, Y. (2014). Visualizing business model evolution with the business model canvas: Concept and tool. In *16th conference on business informatics*. IEEE.

Gassmann, O., Frankenberger, C., et al. (2016). Business Model Navigator, (T. Watanabe, H Morita, Shoei-sha, Trans.).

Hatakama, H. (2018). An AI-based approach to analyze and construct business models. In *Proceedings of the National Conference*. The Japan Society for Management Information. https://doi.org/10.11497/jasmin.2018t10.0_45

Kimata, S., Okada, K. (2019). Study on a methodology to build-up a library of business context mechanisms for advanced business CAD system. In *Proceedings of the National*

Conference. The Japan Society for Management Information. https://doi.org/10.11497/jasmin.201906.0_113

Kizuna, M., Hidema, K., & Okada, K. (2019). Consideration of a method for business model design by employing of business context mechanism. In *Proceedings of the National Conference*. The Japan Society for Management Information. https://doi.org/10.11497/jasmin.201910.0_104

Kizuna, M., Hidema, K., & Okada, K. (2020). Conceptual Design for Business Model CAD system to support business model Design in Scheme Model Projects. *Journal of International Association of P2M, 14*(2). https://doi.org/10.20702/iappmjour.14.2_313. International Association of P2M.

Kizuna, S., & Okada, K. (2019). Study on a methodology to build-up a library of business context mechanisms for advanced business CAD system. In *Proceedings of the National Conference*. The Japan Society for Management Information. https://doi.org/10.11497/jasmin.201906.0_113

Lieder, M., Asif, F. M. A., & Rashid, A. (2017). Towards circular economy implementation: An agent-based simulation approach for business model changes. *Autonomous Agents and Multi-Agent Systems, 31*(6) Springer. https://doi.org/10.1007/s10458-017-9365-9

Minato, N. (2013). Integrated methodology for business model design and evaluation using Business Model Canvas and System Dynamics, System Dynamics, No.12, The Japan Chapter of the System Dynamics Society.

Ministry of Economy, Trade and Industry. (2018). DX Report - Overcoming the IT System “2025 Cliff” and Full-Scale Development of DX. Retrieved September, 13 2021, from https://www.meti.go.jp/shingikai/mono_info_service/digital_transformation/20180907_report.html

Osterwalder, A., & Pignol, Y. (2012). *Business model generation* (R. Koyama, Shoei-sha, Trans.).

Remane, G., Hanelt, A., Tesch, J. F., & Kolbe, L. M. (2017). The business model pattern database - a tool for systematic business model innovation. *International Journal of Innovation Management, 21*(1) World Scientific Publishing Europse Ltd. https://doi.org/10.1142/S1363919617500049

Romero, M. C., & Villalobos, J. (2015). Simulating the business model canvas using system dynamics. In *10th computing Colombian conference*. IEEE. https://doi.org/10.1109/ColumbianCC.2015.7333469

Sakai, H., & Kawai, K. (2005). System dynamics modeling for new product diffusion and its limitations. *Journal of the Research Institute, 31*(3) Momoyama Gakuin University.

Tauscher, K., & Chafac, M. (2016). Supporting business model decisions: A scenario-based simulation approach. *International Journal of Markets and Business Systems, 2*(1) Inderscience.

Vensim HP. (2021). Retrieved September 13, 2021, from https://vensim.com/

Weking, J., Hein, A., Bohm, M., & Krcmar, H. (2020). A hierarchical taxonomy of business model patterns. *Electronic Markets, 30*(3). https://doi.org/10.1007/s12525-018-0322-5

Yoshida, K. (2021). A process for a conceptual design and its simulation toward new business model creation. *Journal of the Japan Society for Management Information, 29*(4). https://doi.org/10.11497/jjasmin.29.4_273

Yoshida, K. (2022). An application of business model patterns and its simulation toward new business model creation. *Journal of the Japan Society for Management Information*. The Japan Society for Management Information. https://doi.org/10.11497/jjasmin.30.4_211

Chapter 5
Virtual Organization, Organizational Intelligence, and Imperfect Information Processing

Tadashi Yamamoto

Abstract After the loss of Eden, human beings have always faced the difficulties of imperfect information processing. No one knows what shall happen tomorrow and the day after tomorrow. And, people strongly wanted to know the future. At that moment, time is starting toward the future. People have rushed into the real world, in which they were suffering from uncertainty and anxiousness.

First, they had thrown themselves on god's mercy. They have trusted their primitive nation's fate and destiny to oracle and prediction in ancient Greek, China and Japan. They offered sacrificial victims to their deities, told their victory and defeat of war, and preyed a fertility of hunting and fruitful crop. Almost simultaneously, people have formed organization to see the far future. They have already known the tactics of war, technology of production, and comfortable life of city.

Organization is, as it were, a castle in the air. Human beings have built it up on the community, before the birth of market, and state. Organization has geometric structure which represents some properties, even though the structure is virtual. The geometric properties generate organizational phenomena and performance.

Organization could have intelligence as well as Human Intelligence and Artificial Intelligence (AI). It has many aspects of intelligence, such as memory, emotion, excellent discernment and inference, and so on. In these characteristics, the core concept of organizational intelligence (OI) is the ability to make a strategic plan for smooth transformation of organizational structure. Adding to it, the core concept consists of setting a goal and predicting organizational future.

Organizational structure has spontaneously grown up and repeated the growing cycle in the birth of enterprise. On the other hand, in any political systems, social systems, and management systems there has continued long-term wave of transformation of organizational structure, only according to technologically growing curve, without intention of organizational intelligence. We can observe that the structure is swinging between hierarchy and decentralized structure.

T. Yamamoto (✉)
Edwin O. Reischauer Institute of Japanese Studies, Harvard University, Cambridge, MA, USA

K. Kijima et al. (eds.), *Systems Research II*, Translational Systems Sciences 27,
https://doi.org/10.1007/978-981-16-9941-2_5

Organizational intelligence is defined as to make a strategy, describe structural design, set a goal, and to predict organizational future. It means that organizational intelligence exists on meta layer in the hierarchy of organizational structure and phenomena. The space of organizational intelligence is out of cybernetic space and out of self-organizing space, but in systems space.

In particular, organizational intelligence makes a plan for organizational structure, while observing and considering the spontaneously structural transformation which can bring optimal transition. Thus, while organizational intelligence is monitoring the transition and simulating the optimal structure using AI, it can support promotion of the metamorphose. This scheme may bring us the newer style of management organization with computerized organizational Intelligence.

Keywords Organizational intelligence · Hierarchy · Decentralized autonomous system · Imperfect information · Technology

5.1 Imperfect Information and Organizational Processing

5.1.1 System Recognition and Meta System

Any system, as is not called as an information system, has always been processing information, while it is working as a functional system for performance. Cybernetics is a model of control systems. It is a black box and universal system applying to creatures, bio-systems, and airplane, according to Norbert Wiener (Wiener 1999). It needs some information to control a system. On the model, input information is perfect for goal seeking. The system with cybernetics proceeds through uncertainty, and reaches to destination.

Cybernetics is one of the imperfect information processing system. Cybernetics never gets oracle. However, it can complete control, using real information and feedback loop. In fact, cybernetics is essentially suitable for control of goal-seeking system and homeostasis. Organizational cybernetics is to control goal-seeking system and I/O system (See Fig. 1.4 in Takahara and Mesarovic 2006). From the other view, this hierarchy shows meta system and meta layers in virtual organization.

5.1.2 Genesis, Oracle, and Prediction

As for a computational model like a decision-making model, if it cannot process imperfect information, and if some necessary information will be added, it will become workable and obtain computability. Adding information not only comes from local scope of horizontal view of decentralized autonomous units, but also comes down from global scope of vertical view of stratification.

When both have some limitation, generating information, such as oracle, enables it to process the information. So, whether they are rational or mysterious, there is some possibility that prediction and oracle enable it to well process the imperfect information. In mythology, a prediction of the three witches and an oracle of god have often saved the life of the hero. Such the prediction is irrational, but it is sometimes rational in decision-making system.

Oracle of god and prediction of goddess are effective to activate the stacking machine. Also, such the divine information makes people be united against a crisis. The virtual information can make people sell or buy stocks, Therefor, it influences on the real stock market. Virtual information also can change the real world. However, original imperfect information remains in social devices.

5.1.3 Imperfect Information Processing

How can some system be processing the imperfect information? If directly we have to deal with imperfect information on the processing, we need oracle and/or prediction. When we consider some model of imperfect information processing, the machine also needs oracle or prediction. It means that the machine itself should generate what information it needs to complete the processing, or acquire any other information which can compensate the lack of necessary information.

What is corresponding to imperfect information in organization is generating information rather than plurality or complexity. Diverging branches on bifurcation can be defined as complexity. However, it is also generating information. AI with machine learning is a typical Imperfect information processing system. This type of AI needs generating information at some step, which enables it to compute the task of image processing. AI needs some information generating on some part, like GAN. But, it consists of stochastic process.

5.1.4 Transforming Imperfect Information

On human history, from the loss of Eden, human beings have always struggled against imperfect information. We did not know the weather of tomorrow for agriculture and fishing. We could not judge whether opening a war would be good or not. The imperfect information struggles have made social devices for imperfect information processing, such as oracle, prediction, market, slavery, social curst, and so on. However, we can understand that they are metamorphoses of imperfect information, and they are not essential solutions of imperfect information.

Market is the most efficient device for processing imperfect information, surely rather than planned economy. However, uncertainty of market is diverging and remains. The slave system and the social curst system are the worst social

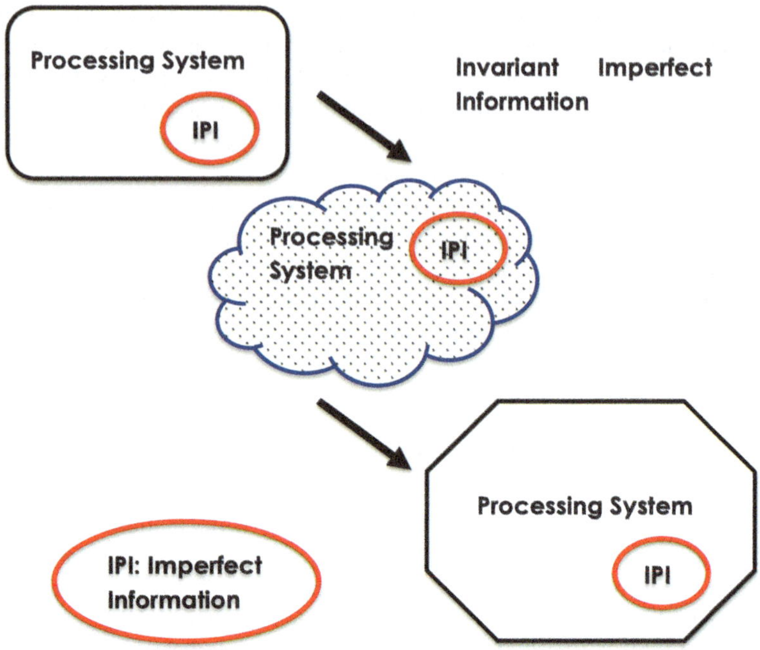

Fig. 5.1 Imperfect information and invariability.This figure is originally illustrated for this chapter

institutions to fix discrimination by means of which a ruler deprives slaves, same human beings, of everything.

5.1.5 Invariant Imperfect Information

These devices show the phases of transformation of virtual organization for imperfect information processing. However, the imperfect information remains as altered other form in each device. So, it is an invariant set for the transformation of organization.

Surely imperfect information remains as an invariant set we cannot solve essentially except pure generating. On the AI, we will face to this problem again. So, it is the only way against essential imperfect information to create enough information to compensate the lack of information. It is to generate a future world enough to provide all necessities. However, it is impossible. So, we need some absolute genesis to generate a new world constantly on virtual world.

This illustrated figure (Fig. 5.1) expresses the existence of invariant imperfect information. Beyond transformation of information processing systems, there remains imperfect information as it changes the form. However, the quantity has never decreased in a new system.

5.2 Middle Level Autonomous Unit in Hierarchical System

5.2.1 Middle Level Autonomous Unit

Whether it is a privately management organization or administrative system, or biological system, if it is hierarchical system, we can easily find a middle level autonomous unit. The middle level autonomous unit (MLAU) has been already given formalized position, but it has informal competences much stronger than it has formally. In human organizations, their position is a middle class manager, such as a branch manager of a bank, a manager of a division of governmental bureau, a director in broadcasting, and a regimental commander of army.

They have intensive competences and gather sufficient information to solve their problems in their layer. It is no exaggeration to say that the business cannot well run without their existence. It is not a post issue, but the ability and informal competences. And, if it is not a formal post, surely such a functional existence will appear in the organization. However, we regard them as an abstract model of MLAU in this section.

5.2.2 Arising MLAUs

Probably, the reason why MLAUs arise in organization depends on local optimal structure. It is the local trap of optimization from which even AI often could not escape. However, in this case, MLAL is the aggregated layer of information and problems. So, if MLAU can process the information and complete the task, it is the better existence in organization temporally.

In Fig. 5.2, as an illustration, MLAU arise on the fourth layer and perform tasks effectively and efficiently, using the information and resources from the bottom of hierarchy and from the top of hierarchy. MLAU has relatively independent power and competences to complete tasks. Usually a hierarchical system is designed as the input comes from bottom and processing is going up. However, MLAU is able to open own channels to same layer of outer organizations. So, organizational power, authorities and competences are informally concentrated on MLAU.

5.2.3 Simulating Transformation of Organization

Using the composite model (See "Composite: A Model of Virtual Organization" in this book.), with some adequate conditions, it is expected that MLAU will remarkably arise in the simulation of transformation of organization. Since the results of simulation of a prototype model, we had the instance data which showed

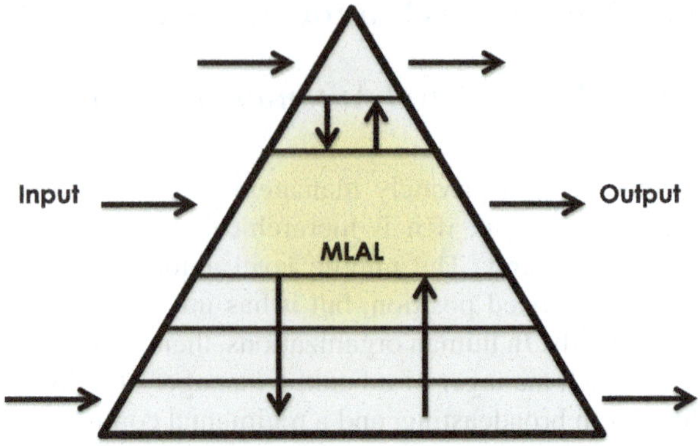

Fig. 5.2 Middle level autonomous layer. This figure is originally illustrated for this chapter

that it was so hard to exceed the forth level of hierarchical height without adjusting the parameter set (Kashiwabara et al. 1993, Yamamoto 1994).

Many of local composites were unable to climb over the fourth layer of hierarchy. From the pilot simulation, we presume that MLAU is one of the phenomena of local optimal structure. If the simulation achieves relatively global convergence, simultaneously the height shall reach seven or eight level. At that time, the fourth or fifth layer is one of transient points. And, it shows that the system escapes from the trap of local optimal structure. If the level of layer stops at the fourth or fifth layer, it can show existence of Middle Autonomous Layer which contains MLAUs. Even though it shows the drop into local optimal trap, sometimes it can be useful for organizations under some conditions. It is well worth consideration that there occurs the vertical distribution of power and information in stratification.

5.2.4 Proposition of MLAU and Global Optimality

From the results of the pilot simulation and fact finding of MLAU in virtual organizations, we can obtain one proposition. If the summation of performances produced by each local optimal unit is sufficiently bigger than performance produced by a global optimal unit, and if the summation of the computational complexity of local optimal units is equal to or smaller than the computational complexity of a global optimal unit, then the organization containing many of the local optimal units is better than the organization containing one global optimal unit. This is the problem of compering quantity of performance to computational complexity.

In realistic, if the summation of performances produced by each local optimal unit is equal to or relatively bigger than performance produced by a global optimal unit, and if the summation of the computational complexity of local optimal units is smaller than the computational complexity of a global optimal unit, then the organization containing many of the local optimal units is better than the organization containing one global optimal.

In more realistic, if the summation of performances produced by each local optimal unit is equal to or relatively smaller than performance produced by a global optimal unit, and if the computational complexity of a global optimal unit is sufficiently bigger than the summation of the computational complexity of local optimal units, then the organization containing many of the local optimal units is as good as the organization containing one global optimal.

The organization containing many of the local optimal units means containing of MLAU. So, considering the computational complexity of a global optimal unit, a management organization with MLAU is pragmatically superior to uniform hierarchy.

On the composite model, "unit" is changed to "composite" (See Sect. 5.3.3). Surely, the basic condition set which determines the performances and computational complexities depend on the transformation of system structure and the growing curve of technology. On the job load, algorithm selection influences on computational complexity.

In the next stage to MLAU organization, advancing the computer technology, the new age of uniformed hierarchy will come to management organization, when the computational complexity of a global optimal unit comes to be sufficiently smaller than the summation of the computational complexity of local optimal units.

5.3 Organizational Intelligence

5.3.1 What Is Organizational Intelligence?

Late Professor Takehiko Matsuda originally advocated the theory of Organizational Intelligence in management organization. At the dawn of the AI age, it was a simple analogy of machine intelligence and human intelligence. The machine intelligence is now called Artificial Intelligence. He thought that there would arise own intelligence in organization, in particular in management organization. On the first idea, Organizational Intelligence is thought to have memory, emotion, own decision-making process, and so on.

Surely, we can see some kind of intelligent characters on an organization, such as common memory, own standards of decision-making, shared emotion among the staffs, and community songs. From the historical view of management, there developed small devices which characterized organizational culture, such as recreational activities, a company athletic meet, a baseball tournament, and a home party. Sometimes, they bring a big family style into a company.

Organizational Intelligence is emerging beyond group intelligence and summation of individual intelligence of members. As we investigate the characteristics of a model of organizational intelligence, we will set the two directions in this research. One is an abstract model of virtual organization. It is a model apart from real organization. The other is a computationally neural-scientific model.

Organizational Intelligence has an ability to predict organizational future. It can observe bifurcation and image possible future. Simultaneously it can recognize spontaneous transformation of organizational structure, and derive global information processing. Organizational intelligence is, as it were, a bridge between operational space and spontaneous phenomenon.

Classic organization is often designed according to famous line-staff system as well as military systems. This is a typical designed hierarchical system, not spontaneous stratification. Even though it is very solid structure in strong hierarchical system, substantially Middle Level Autonomous Layer (MLAL) and its units (MLAU) arise and play an effective role through the management process. It is one of the characteristics of organizational intelligence, because it can allow the units to act freely in the organization.

5.3.2 Virtual Organization and Organizational Intelligence

In this short article, it is impossible to discuss the comprehensive characteristics of general organization. Thus, we try to grasp the core of the characteristics. Organization is composed of sub-systems and contains many local processes of task processing. And, it can process imperfect information with gathered and condensed information, data, and technology. If so, it is the typical characteristics of organization that can settle on a long-term strategy for far future, formulate a strategic plan against uncertainty of outer environments.

Now, extreme progressing of telework makes virtual organization rise up on internet services. It is combined to real management organization. Each staff is connected to others on the network and the opportunities to meet each other is so rapidly reducing that they are unable to feel the tie and royalty. Surely, regionally local management communities may sprout out anywhere on network. It makes organizational intelligence change to more virtual and functional existence, and local network intelligence occurs.

Organizational intelligence emerges from organizational structure and process. The main role of organizational intelligence is to make a strategy, design organizational structure and process, and to set a goal, while it is observing the transformation of organizational structure. Also, prediction of organizational future is the important role of organizational intelligence (See Fig. 5.3).

Thus, organizational intelligence exists in the meta layer rather than cybernetics, self-organizing system, goal-seeking system in the meta hierarchy of organization (See Fig. 1.4 in Takahara and Mesarovic 2006). In this meaning, organizational intelligence is out of the space of organizational cybernetics (See Fig. 1.3 in Takahara and Mesarovic 2006).

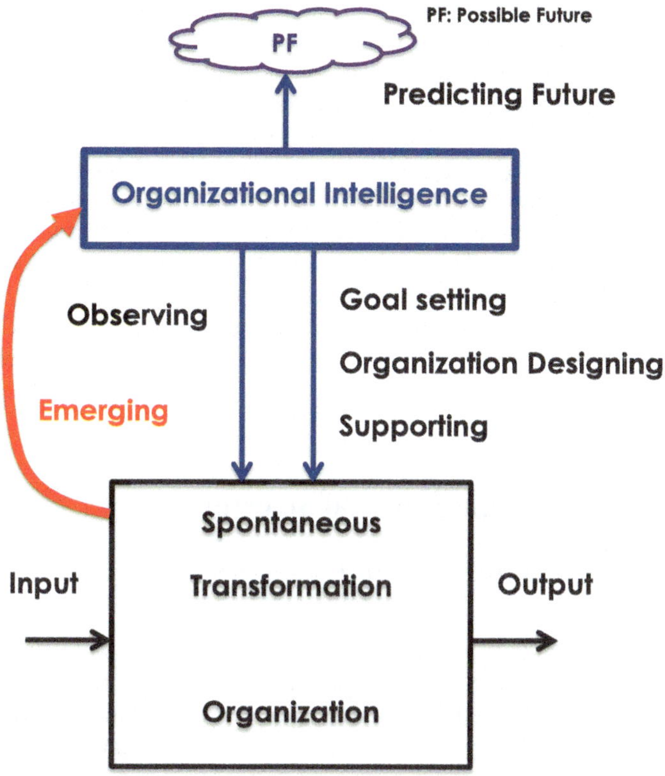

Fig. 5.3 Organizational intelligence. This figure is originally illustrated for this chapter

5.3.3 Composite Model and Organizational Intelligence

If we exclude the strength of a tradition, solidarity, royalty, folklore, memory of athletic meet from organizational Intelligence, there will remain essential organizational intelligence such as planning strategy, changing structure by itself, choosing right direction to the future in bifurcation, and decision-making under complex environments. Therefore, organizational Intelligence is regarded as collect intelligence with emergency beyond the connective relations. So, a composite model seems to be useful for modeling organizational Intelligence.

If we try to apply the characteristics of composite to such functions as planning and performing of strategy, only higher-level composite can play the roles with the global scope. Changing structure by itself is beyond the genuine characteristics of self-organizing. Choosing right direction to the future in bifurcation is also the role of higher-level composite with the representation of bifurcation. And well decision-making under complex environments depends on optimal structure by the transformation.

Composite is a model of virtual organization based on decentralized autonomous units. In the composite model, virtual stratification arises, in accordance with the composing frame and the local rule sets of the units. Composite and its nested stratification are fluctuating widely, while they are processing information and seeking for optimal structure to perform their task (Yamamoto et al. 1992).

According to the composite model, organizational intelligence will arise from decentralized autonomous units, too. As well as composing process, local rule sets will play a decisive role. Composite model shows spontaneous stratification and higher-level composite. This higher-level composite is related to prediction of organization to the far future. It is a role of organizational intelligence. On the other side, it is also an analogy of the developmental process of human intelligence and psychology. Simulating organizational intelligence is the important tool for these investigations.

5.4 Geometry of Organizational Structure

5.4.1 Transformation Between HS and DCAS

As historically and widely organizational phenomena from biological system to political system, hierarchical system (structure) and decentralized autonomous system (structure) have looked like alternating reciprocally (Yamamoto et al. 1993). Why has it occurred through human history? In fact, we can easily observe an integrated hierarchical system (structure) or a decentralized autonomous system (structure) in any organization, whether it is in nature or artificial. Also, there appears some mixed phase among them.

All systems and organizations are based on decentralized autonomous system and its units. On the surface, hierarchical system virtually arises up from decentralized autonomous units. And, it collapses and goes back to decentralized autonomous system. So, this organizational dynamics of composition and decomposition produces the transformation between hierarchical structure and decentralized autonomous structure as a geometric structure of system.

5.4.2 Growth of Technology and Transformation

Proposition is following. As technology is growing up, the system structure would alternate hierarchical system and decentralized autonomous system. If now it is hierarchical system, and technology is growing up, the structure will change to decentralized autonomous system. Conversely, if now it is decentralized autonomous system, and technology is growing up, the structure will change to hierarchical system. We built the model of alternation and gave the mathematical proof (Yamamoto et al. 1993, Yamamoto 1994).

Hierarchical system and decentralized autonomous system are reciprocally transformed by development of technology. It is the advancing technology that gives the reciprocal transformation, not the declining of technology. So, if some technology will advance on the hierarchical system, it must be transformed to decentralized autonomous system. Next advancing technology in the decentralized autonomous system, it shall change to hierarchical system. On the same system, as continuously advancing technology, not declining, same system shows hierarchical structure or decentralized autonomous structure reciprocally.

Figure 5.4 expresses the relation between technological growth and transformation of structure illustratively. The technology growth curve is going up, which shows monotone increasing. At that time, structure of the system or the organizational structure is constantly and reciprocally fluctuating between integrated hierarchical structure and decentralized autonomous structure. Considering local stability of the system structures, the transformation curve represents the stairs shape. In fact, there occurs a time rag between technological growth and system changing. In the Fig. 5.4, IHS is Integrated Hierarchical System (Structure). DCAS is Decentralized Autonomous System (Structure).

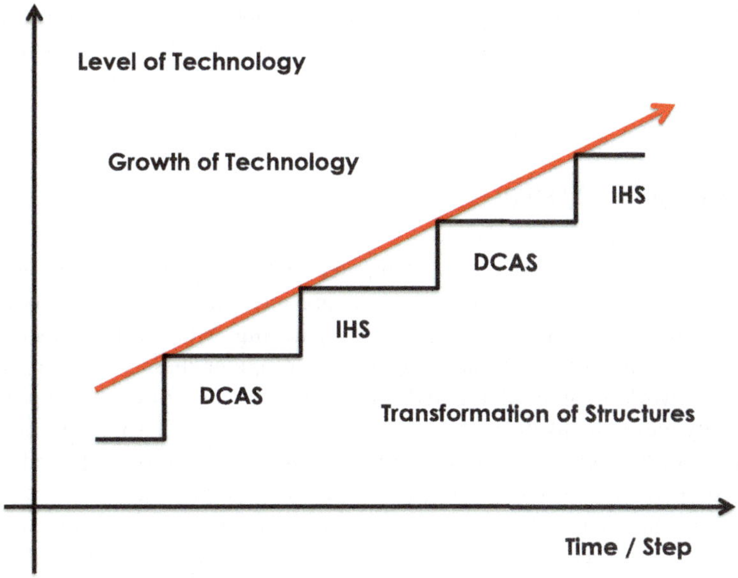

Fig. 5.4 Stairs of IHS and DCAS. This figure is originally illustrated for this chapter

5.4.3 Optimal, Efficient, and Advantage Process

We can observe the phenomenon of the structural transformation in any social systems, organizations, and natural systems. The technology is neither necessarily scientific technology nor engineering technology. It is frequently industrial technology and social technology. So, it is not right that hierarchical system ought to be efficient by its structural effects if we design it under any situation, and that we would design hierarchical system in order to obtain efficiency and advantage.

When we stand on decentralized autonomous system, there happens same situation by same reason. Thus, as long as our civilization is on advancing, any social systems and organizations will show the phase between hierarchical system and decentralized autonomous system.

When some hierarchical system changes to decentralized autonomous system, a historian is apt to say that it really shows degeneration and collapsing of the civilization, and that it is on declining. However, back of the transformation, historical technology has so rapidly progressed that it made remarkably structural dynamics.

This theory suggested that any social systems are unable to escape from the law, and that the alternation obeys the growing up of technology. So, policy maker should not design distributed structure when the technology does not grow up yet, Industrial policy, too. While a hierarchical system or an integrated system is well working and completes efficiency, if the system designer, policy maker, or organization designer may try to set a distributed system, surely the system will lose the efficiency and advantages of technology.

5.4.4 Computer System

On the historical trajectory of the computer technology, the age of integrated systems and the age of distributed systems are alternating, as the computer and network technology has been continuously advancing. Moreover, we can observe the same phenomenon from the deep inside of a computer chassis to the topological surface of network. In an implemented computer CPU, it is composed of physically and logically hierarchical structure, integrated systems, and distributed topology of processing. Such a type of CPU needs the advanced technology of prediction in order to complete the more efficient order processing in this small world.

As well as the phenomenon of the computer structure, social systems have had same tendency to alternate the two phase of systems structure. The eternal problem of political institution which a central integrated system or decentralized distributed system is better depends not only on the policy making but also on the social and information technology, if we may seek for social efficiency.

5.4.5 Political System

In this case, political power generates political systems and decides what the type of structures is, such as a tyranny system as an integrated hierarchical regime, feudalism as a middle level distributed autonomous system, absolute monarchy of the Bourbon as a mixed phase. It is profoundly interesting that even the Bourbon absolute monarchy has contained the phase of decentralized distributed systems. So, Bourbon de France needed delicious French cuisine and famous wins for lavishing hospitality to great dukes and admirals at the Versailles.

Therefore, at the time of designing large scale public system such as local autonomous institution of state, we should observe the present state and tendency of technological advancement very carefully.

5.4.6 Production System

If some bureaucrat, politician, CEO, and other high-ranked decision-maker would set wrong structural design on production systems or firm systems against the spontaneous dynamics of technology, it should court disaster to the state, economy, company, and any other organizations.

When integrated and hierarchical production system is the most efficiency as advanced technologies, introducing decentralized and small production system must bring low productivity and environmental pollution. It is an important problem on policy making. Historically, many dictators and bureaucrats misunderstood and failed it.

From the modern view, it is the symbol of civilization that the blast furnace lifts its head into the sky, and that the integrated nuclear power plant provides electric power to several megalopolices. This phase is temporally stable and sometime transformed to next stage of decentralization. However, before the timing, if someone tries to change the structure against growth of technology, it brings serious disaster.

Small business, called as venture business, can be successful, when it catches the time when the tendency of advanced technology is turning into decentralized autonomous phase from integrated hierarchy phase. The secret of venture's success basically depends on the decentralized autonomous circumstance in which anyone can start the small business with high technologies and grow it to bigger day by day from the smallest sprout.

5.4.7 Power and Generating Organizational Structure

Geometric structure of organization and social systems depend not only on information or imperfect information, but also on power, something like energy and any undefined quantities. We can easily find the typical political structure after the big war, revolution, and fiercely political struggles. At that time, tyranny centralized system appeared.

On the other hand, if generation power were weak, mixed phase between centralized system and decentralized system would appear. Such the phenomenon is so universal that we can observe it from historical polity to private organization. These geometric structures are based on power rather than information.

So, in this political case, power as political quantity decides type of political system. We have no good measure for such the quantity. However, we can judge the amount from the appearance of geometry after generation. Strong power generates absolute monarchy, weak power makes fragilely distributed institute such as feudalism. Surely, we can observe the middle level appearance between centralized integrated system and decentralized distributed system such as Ottoman Turkey Empire, Japanese Edo polity and many of democratic polity.

Organizational structures should be designed according to a measure of some structural quantity which may generate it naturally. If the design ignores this idea, the organization could not keep good characteristics of efficiency and abilities to perform task in the field.

5.4.8 City and Integration

This idea can be easily applied to the problems of cities and local autonomy. Both the problem of distribution of cities and balance of local autonomy should obey the frame of alternation of systems structure first, rather than the graph theory. Inevitably, it is the autonomous city that is generated from some virtually structural quantity. And, the autonomous cities make the network as a decentralized autonomous system.

Roman city is one of the devices of the Roman Empire in order to conquer a frontier, govern the area, and extend the imperial prestige to all sides. It was designed, normalized, and packaged. At the age of the republic of Rome, there was a forum and a coliseum. After establishment of Catholicism as the state religion, a cathedral was built in the center of an Episcopal city. The rise and fall of a Roman city also obeyed the law of transformation.

5.5 Conclusion

Organizational structure shows its geometric prosperities and spontaneous transformation, which generates organizational phenomena and characteristics. Organizational intelligence emerges from organizational structure and process. The main role of organizational intelligence is to make a strategy, design organizational structure and process by itself, and to set a goal. Moreover, organizational intelligence can see the far future of organization.

So, organizational intelligence exists in the meta layer rather than both cybernetics and self-organizing system in the meta hierarchy of organization. In this meaning, organizational intelligence is out of the space of organizational cybernetics.

Thus, it is organizational intelligence that should observe the spontaneous behaviors of organization and form optimal structure of organization. In particular, organizational structure is swinging between integrated hierarchical structure (system) and decentralized structure (system), according to growth of technology. Spontaneous transformation of organizational structure is optimal process under the conditions. So, organizational intelligence should obey this law and structural phenomenon, and make a strategic decision by itself.

Appearing of Middle Level Autonomous Unit (MLAU) is also one of the typical organizational phenomena. It is spontaneously made up and confirmed. When we try to apply composite model to this MLMU phenomenon, we can obtain one result that MLMU is a set of local optimal points, using a result of the computer simulation. So, this trial combines organizational intelligence with AI. Thus, next trial is to build virtual organization with organizational intelligence based on decentralized autonomous units on network, using AI technology.

Acknowledgements Late Professor Takehiko Matsuda has always told us a lambent wit with his intense face. What he told us was very suggestive, such as memory of Herbert Simon, academic life of America, and establishment of Information Management sciences. One day, he talked of whether it might be organizational intelligence, intelligent organization, or intelligence organization. His wit was a spring of creation and a window open to the world.

Professor Yasuhiko Takahara has always climbed long stairs in a tall building, not using an elevator, and often shut himself up in a computer room until his retirement. Someday, I tried to climb the stairs with him. It was so hard and we were talking on climbing mountains. From his positive attitude and high standard of evaluation, we learnt so many things as well as his brilliant research results.

References

Kashiwabara, K., Yamamoto, T., & Nakano, B. (1993). An algorithm for generating artificial society and the simulation model of artificial creatures society. In *Proceedings: The 4-th autonomous distributed system symposium* (pp. 125–130).

Takahara, Y., Mesarovic, M.D. (2006). Organizational Structure: Cybernetic Systems Foundation. Private edition.

Wiener, N. (1999). *Cybernetics: Or control and communication in the animal and the machine*. The MIT Press.

Yamamoto T., (1994). *A study on complex distributed composite and self-generating - Process of complexity and emergence on social phenomena and living phenomena*. Doctoral Thesis, Tokyo Institute of Technology, Oookayama, Tokyo.

Yamamoto, T., Iijima, J., & Nakano, B. (1993). A trial of the foundations of hierarchical structure and decentralized autonomous system. *Journal of the Japan Society for Management Information, 1*(3), 75–90.

Yamamoto, T., Nakano, B., & Matsuda, T. (1992). System, information and self-dynamics. In *International conference on economics / management and information technology (CEMIT92/CECOIA3), proceedings* (pp. 211–214).

Chapter 6
Composite: A Model of Virtual Organization

Tadashi Yamamoto

Abstract Organization is, as it were, a temporally stable whirlpool appearing on flow of imperfect information processing. At that time, there are going on two processes concurrently. One is to make optimal organization to perform task. The other is to process imperfect information. If we describe this process as an inherent model based on discrete structure, many decentralized autonomous units continuously try to connect to others, from which some nested hierarchical system forms with optimal structure.

Composite is a model of virtual organization. It is also a model of hierarchical system, based on distributed autonomous units and their relations. Each unit has own basic endowment sets and operators. It is trying to connect to each other, using its local scope under some local rules, and makes network topology. A composite is hierarchically composed of the nested sub-composites through this continuous process. It is spontaneously growing up by composition, and sometime later collapsing by decomposition.

Composite and its composing process show that there arises characteristic geometry and structural complexity. At the base layer, Decentralized Autonomous Units (DCAUs) and their connecting form some topological geometry, which shows some kind of complexity. Also, spontaneous stratification with nested structure represents generating complexity. In the middle hierarchical layer, there will emerge measurable complexity. On the other hand, composite is an information processing system simultaneously with a self-organizing system. In particular, it deals with imperfect information. So, it makes bifurcation and a new branch for future forecasting. Higher stratified composite based on wider domain of DAUs could see the further future in the global scope.

The top of composite is transformed, according to composing process. It can be optimal process with global scope. In simulation, there often happen many local optimal solutions by local scope. As it can procure the higher global scope, it will escape from a trap of local optimal structure. Composite holds nested stratification and composing process with repeating operation and recursive operation. This

T. Yamamoto (✉)
Edwin O. Reischauer Institute of Japanese Studies, Harvard University, Cambridge, MA, USA

K. Kijima et al. (eds.), *Systems Research II*, Translational Systems Sciences 27,
https://doi.org/10.1007/978-981-16-9941-2_6

composing process generates complexity and computational complexity at the conceptual simulation.

From the Paradise Lost, human beings have always developed the newer systems for imperfect information processing. However, there remains invariant imperfect information against the transformation of the systems. Thus, it is necessary for any information systems from CPU to AI, and to world market to predict the near future. While they are seeking for efficiency or advantages, surely imperfect information will appear in their small spaces. Composite model enables us to generate inherent information for predicting the near future against imperfect information. However, for acquiring genuine generator by means of which we can overcome the invariant part of imperfect information, we need to develop a conceptual model of Transcendental Genesis.

Keywords Bifurcation · Complexity · Imperfect information · Stratification · Prediction

6.1 Introduction

6.1.1 System Recognition

System finding is a very interesting and intellectual search in nature and social phenomena. Norbert Wiener found out his famous cybernetics from observing bird flying (Wiener, 1948/1999). How to fly to final destination of a bird became the concept of control system. In that time, cybernetic system is a meta system recognized as a black box. So, it is characterized by abstract, universal, and anonymous properties.

These conceptual properties should have formed the core concept of artificial science. Designing a concept, building the logically operational model, and simulating phenomenon and acquiring data, artificial sciences are due to consist of these steps in implementation process. From 1930s to 1950s, Cybernetics, computer technology, DNA and biotechnology, and information entropy theory have been discovered following one another. These new results have come to be foundation of the new sciences and technologies in the twentieth century.

At the same time, development of systems science has been rapidly proceeding. I/O system shows the causality and state transition, Turing machine is a first model of computation. Cybernetics brings control theory and stability. Cellar automaton generates global patterns from local rules and relational activities of distributed cells. It is a typical idea of discrete mathematics which shows stronger representation than continuous mathematics. And then, next interests have focused on the concept of self-organization. It is so interested for us how to form the system and organization from disorder chaos, and discrete bottoms.

In particular, Decentralized Autonomous System (DCAS) can be a universal base of system arising. As DCAS transforms into variety of structures, according to the local rule of each autonomous units, the global and higher dimensional structure will arise and represent global and emergent characteristics of the whole system. Stratification is also based on DCAS.

6.1.2 Geometry and Structure

Basically, generation of systems and their behaviors can form inherently geometric structure which will exhibit characteristic phenomena. Systems phenomena will be exhibited by its geometric structure. Also, it can bring emergent and functional characteristics onto the system. Surely, DCAS shows plurality, diversity, and elasticity. Hierarchical system shows efficiency, high performance, and vertical rectification of information. Fractal and foliation may make some complex and chaotic phenomena appear on the system.

It is too many to enumerate the relationship between structure and phenomena or functional characteristics presented on the surface of the system. In particular, trajectory of flows on process will make the structure, such as network topology, active neural network, brain neurons, and circulation of money and distribution of commodities. Biological systems finding gave us strong impression on the new modeling of complex system. Immune system is composed of many functional components, such as T cell, B cell, and antigens and antibodies. The immune model seemed to be useful for simulating artificial intelligence and societies.

6.1.3 Hierarchy and Stratification

Hierarchical system has hierarchical structure containing multi-layers. It is frequently designed as a hard system, whereas stratification is spontaneously stratified and composed of many components of subsystems. This composing process some tome may form hierarchical structure from the bottom of decentralized autonomous units (DAUs). Hierarchical system is often fixed, and has stable and solid structure. Stratification can change the structure and transform into the next style spontaneously.

6.1.4 Complex System and Chaos

After the boom of self-organization in the academic field, complex system and chaos were in the spotlight again, because both complexity and chaos were expected to emerge something useful for problem solving. Evaluation of complex and chaos changed to the goodies from baddies. On the back of this turning of eyes, there is the problem of imperfect information which remained implicitly in deep tissues of every system models and information processing. Whatever device might be developed to process imperfect information, it would arise from the hideaway. In other words, imperfect information is an invariant set through any transformations of the system. So, complexity and chaos are expected to bring both emergent qualities and quantities against imperfect information.

6.1.5 Hierarchy and Complexity

As Herbert Simon also pointed out, hierarchical structure is related to complexity. His hierarchic system is composed of interrelated subsystems (Simon, 1999, p. 124). It is the fact that this type of composed systems has the possibility to show complexity. Surely, we need a model of "systems in which the relations among subsystems are more complex than in the formal organizational hierarchy" (Simon, 1999, p. 125). Also, geometric characteristics of hierarchical structure explicitly show complexity.

On the contrary, complex model often contains hidden hierarchical structure. Mathematical modeling reveals the hierarchy related to complexity. When we solve Kdv equation of Soliton which shows complexity and the chaotic phenomenon of wave, we can find some hierarchical system which lurks in the way of the solution. We should consider both explicit phenomena and implicit structure on complexity.

6.1.6 Vertical Section of Hierarchical Structure and Comprehensive Recognition

Considering the vertical section of hierarchical structure, we can obtain comprehensive recognition of hierarchical system. From a quark to the galaxy, from a cell to the whole body and conscious, and from an individual to the world, there arise hierarchical structures. So we need the comprehensive recognition in order to build a simple model, operate computer equipment for monitoring the situation, and to create the future prediction. In this meaning, formalization of a hierarchical system as an abstract system is so significant works.

As Herbert Simon wrote, biological system is composed of "well-defined subsystems" (Simon, 1999, p. 186). Bio-system is a typical stratified hierarchical system with nested structure. And, it is necessary to use both vertical and horizontal systems recognition in order to comprehend the whole biological phenomena from DNA to body conscious and psychology. Immune system has complex mechanism and implicit properties through its activation process.

Swarm of animals, birds, and insects behaves as if it were one organized body. So, we can regard it as instances of self-organization and emergence. A modeling of living creatures from immune system to swarm intelligence is expected to expand the view of systems recognition to the area of self-organization in the 1980s at the latest and to be useful for technology of artificial intelligence in the 2010s. Moreover, biological phenomena of generation and annihilation are expected to suggest the core ideas of the next step.

6.1.7 *After Complexity and Chaos*

In any systems whose function can be regarded as information processing, implicated imperfect information has remained and becomes a malignant guitar neck of logical processing from market forecast to AI prediction. Recently, a variety of AI methodologies are realized immediately by proceeding of computer power. Whether AI as is now implemented is going to right distance or not, the core of AI technologies is so deeply related to imperfect information that it may need generating new information.

Controllable complexity and expected chaos have some effectiveness to solve these problems, after the boom of self-organization. However, it is not essential solution. In consequence, it is necessary for us to consider transcendental genesis which provides absolute addition of new information, since oracle and prediction of Greek gods.

This article is basically edited from (Yamamoto, 1994a) with additions.

6.2 Composite

6.2.1 *Problems*

What kind of process will make virtual entities be rising up on the network? In concurrence with it, how can the system be processing imperfect information, while it is transforming the structure by itself? It should happen on the field of distributed autonomous units (DAUs). So, how do DAUs spontaneously make some integrated system, global structure, and optimal organization? Each DAU is a small entity which holds characteristic operation and relates to each other by networking. So, How many DAUs can be connected and stratify a higher dimensional hierarchical system? Also, to what extent will the stratification rise up as a hierarchy?

The basic concept of the DAUs is a natural assumption for individual human beings dispersed over social field and artificial entities on computer network. And, we need spontaneously generated stratification in the model, neither designed nor controlled. What fulfills the conditions is called as composite. Through building a conceptual and operation model of composite, we can think about complexity and emergent qualities which may occur on composite. Also, through simulating the model, we can investigate into the properties of virtual organization.

6.2.2 Composite: A Model of Virtual Organization

Composite is a model of virtual organization. It is precisely a model of spontaneous stratification and nested hierarchical system, based on decentralized autonomous units (DAUs) and their relations. Each unit has own property set of endowments and operators for information processing and networking. DAU is trying to connect each other, using an individual scope.

A composite is composed of DAUs and stratified with nested composition. Surely, it is spontaneously growing up and sometime later collapsing, which shows characteristics of the transformation by composition and decomposition. Composite shows global properties and emergent qualities from recursive and repeating process of simple calculation. Building the mathematical model of composite, we can obtain several analytical results, and implement the algorithm for simulating artificial societies and virtual organizations.

Composite is virtual existence, and the stratification consists of meta layers. It has virtual functions and meta operations. Virtual composite on meta layer can operate lower level composites and real parts of AU. Composite is not only bottom up and nested, but also acquires an essential ability to operate them. Whole system of composite and stratification also shows features of multi-layer meta system.

6.2.3 Frame of Composite

The concept of Composite had a simple frame. The frame of composite has universality. First, it is based on many of Decentralized Autonomous Units (DAUs). Thus, the real part of composite is the Decentralized Autonomous System (DAS) which consists of sufficiently many n-units. Each Decentralized Autonomous Unit (DAU) has own characteristic endowments and autonomous operators. Also, each unit holds own scope to search another units and their endowments, and relational operators. The operation model of DAUs is defined as the local rule.

Second, each composite connects each other and composes a new composite with higher dimension, level, and rank. In this expression, dimension is a measurement in composite, level is the number of layers in stratification and hierarchy, rank is the number of orders in transformation. Higher level composite consists of lower level composite. Thus, composite has nested structure.

Third, composing process contains recursive and repeating operators which evaluate every component composite. Forth, each composite has each scope as well as a DAU does. The range of the scope of a composite is, at least, equal to the range of the aggregation of scopes endowed by each contained composite.

Fifth, the field of DAUs is open. It means that there are sufficiently many DAUs and invisible DAUs from the scope of a DAU (DAU1). If the invisible DAU (DAU iv) is encountering, because the DAU (DAU iv) can see the DAU (DAU1), the DAU (DAU1) shall recalculate the composing process. At that time, encountering of the

DAU (DAU iv) enables the DAU (DAU1) to catch the DAU (DAU iv) in the scope. Every composite obeys the same rule and procedure vertically and horizontally.

6.2.4 Local Rule Set

Each Decentralized Autonomous Unit (DAU) acts under the local rule set. A local rule set is very simple. Each DAU is endowed with own properties and requisites to process the information, to solve the problems, and to perform the tasks. Matching the properties to the requisites, if it is not enough to complete performance, each unit searches for another units in order to exchange its properties to requisites, using own scope. And, when the unit finds out the best partners who agree to the exchange, these units are connected and compose the first order composite. This local rule is one of the typical exchange models and contains imperfect information processing. Imperfect information becomes motivation of connecting another units.

In this model, each endowment is variable according to change of input. The inner operations and endowment are anything at all. We can use a different local rule set, too. According to the composite frame, based on the local rule set, a composite is composed of DAUs, and repeating the composing process. Many composites with high dimension, level, and rank, will arise, transform, and annihilate virtually in the composite field. At last, Mathematical modeling and simulating the composite bring us interesting results.

6.2.5 Decentralized Autonomous Unit and Scope

At the base layer, sufficiently many autonomous units are decentralized and dispersed on the plane. And, any entity does not encompass the group of DAUs as a whole and unified entity. Each independent DAU is trying to connect each other, using its own scope, operators and endowments. In Fig. 6.1, an AU (DAU) (left) is connected to another AU (right), using its scope, after evaluation of the properties and requirements of endowments. The right side AU cannot come within the range of the scope in this plane.

Figure 6.1 is originally illustrated for this chapter based on Yamamoto (1994a).

In this model, we set exchange rule into Decentralized Autonomous Unit (DAU) in order to drive the simulation model. In the Fig. 6.1, each AU (same as DAU) can see endowments of another AU in the scope. If the AU (left) needs some elements in the endowments of another AU (right) in order to work completely or to process imperfect information, AU (right) also needs some elements in the endowment of another AU (left), the exchange conditions are satisfied and the two DAUs will make connectivity.

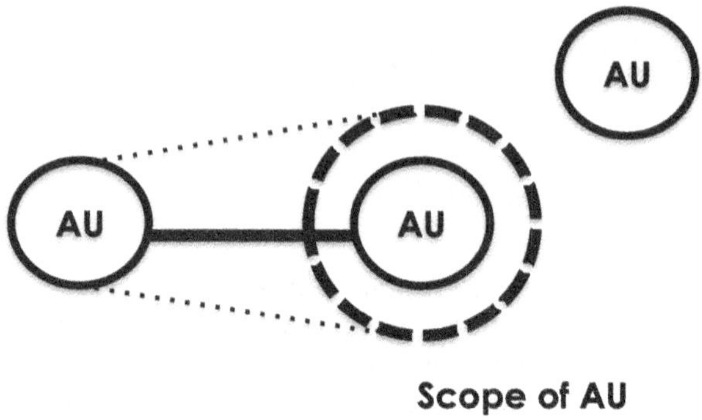

Fig. 6.1 AU connectivity and scope

6.2.6 Local Scope and Highly Global Scope

Each AU has own local scope by means of which it can see another units and their endowments. Each local scope has some limit to see another units. It should be some restriction. In accordance with composing and stratifying, it will be growing up to higher global scope which enables the units to see another units in the area they cannot see by themselves. The higher composite unit can see another units in the same layer.

It means that the higher composites contain lower composites which can see another invisible composites by their own scope. Because, trough the stratification and the recursive operation of composing, the bottom units come to be able to use the higher global scope. In this sense, connecting of DAUs can play a role of a counter plan to solve imperfect information problems. From the summation (union) law of scope, global scope is, at least, equal to the summation of local scopes of the contained DAUs.

The first level composite is composed of these two AUs and connectivity. Simultaneously, the composite is defined as a new scope, endowments, and operators. In Fig. 6.2, the scope is wider than scope of AU, because it contains the scope of AU (center). Thus, now AU (left) can see every AU through the scope of the composite. If the scope contains an AU which is not contained in the union of the two scopes, it is some emergent character, we think.

Figure 6.2 is originally illustrated for this chapter based on Yamamoto (1994a).

6.2.7 Composing Process and Stratification

Composite requires (1) Based on decentralized autonomous unit (DAU), (2) Nested vertical relation, (3) Recursive and repeating operations for composing, (4) Scope and summation (union) law of scopes. Adding to them, a set of local rules of a unit

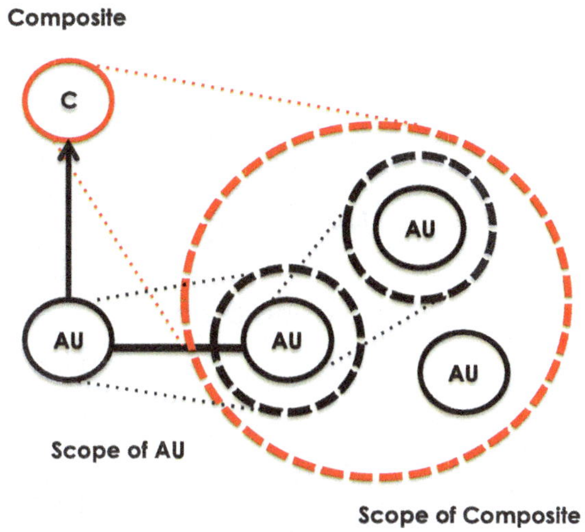

Fig. 6.2 Composite and scope

makes stratification and other geometry of structures. So, there arise nested structure and virtual composition on the field of decentralized autonomous units.

A composite in the higher level (layer) contains composites in the lower level as a component. The higher level composite is connected to the same level composites in the plane. And, they will compose next level composite, evaluating each other horizontally and vertically trough the composing process. Thus, these recursive operations make nested stratification. The generated composite is a virtual entity which arises on the distributed decentralized autonomous system.

Figure 6.3 shows comprehensive structure of composite and composing process as an illustration. In the Fig. 6.3, at the base layer, AUs are distributed and connected, using their scope and connecting operators. It makes the first composite (green line). After then, in each layer horizontally and among layers vertically, same recursive operations are repeated. It makes the higher composite (purple line) as a top of composite.

The horizontal line shows plurality of AUs and their dispersion. It is usually described as dimension, because AU is a base of composing. Moreover, Each AU, and each composites, can make summation, or more. Each AU has own scope, composite, too. Each scope can make summation, or more. And, if we assume some quantity of state of composite as CQ (C), it makes summation, or more, for each composite on the plane.

The vertical line shows height of stratification. It is usually described as level or rank. Vertical repeating of recursive operations can be described as production and factorial times through compositing process. Thus, composite frame and composing process show horizontally summation and vertically factorial at least. Number of AUs or composite is usually finite.

Figure 6.3 is re-illustrated for this chapter based on Yamamoto (1994a).

Figure 6.4 is originally illustrated for this chapter.

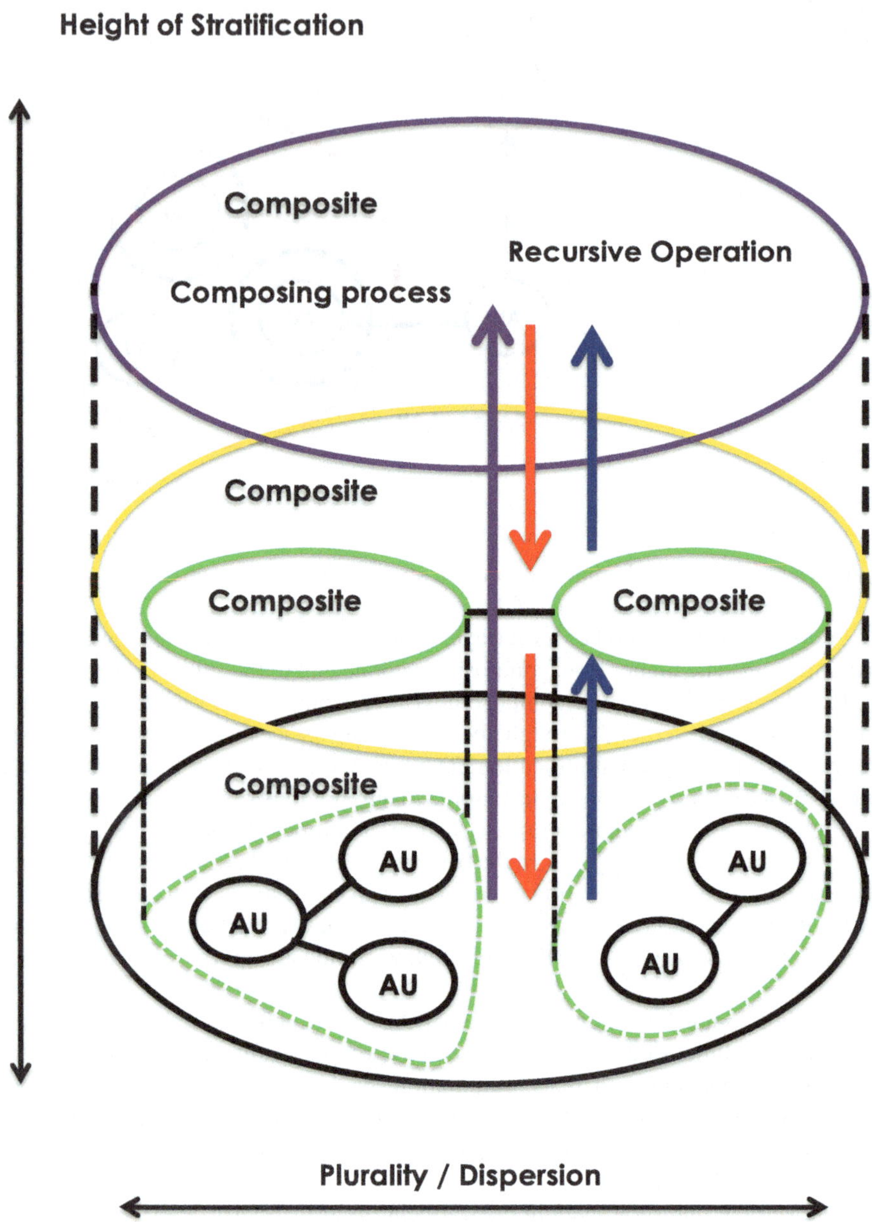

Fig. 6.3 Composite

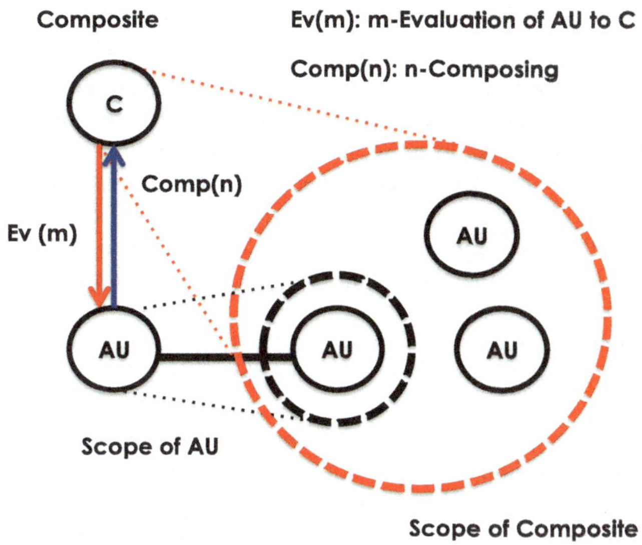

Fig. 6.4 Composite and recursive operation

6.2.8 Recursive Operation and Composing

Composing process contains recursive operations to evaluate composites and to judge whether each composite is suitable for connecting or not. And, it forms a higher composite with nested stratification which contains itself. Basic operations are repeated vertically between higher units and lower units. In other words, the repeated and recursive operations aim to reach to global convergence, escaping from local optimal solutions. This repeated process makes some complexity. An autonomous unit (AU) has its operators and endowments. The composing process changes the characteristics of the operator.

In the Fig. 6.4, the illustration shows that there are activated repeating and recursive operations between C and AU. We can exchange AU for C of lower level. Figure 6.4 is originally illustrated for this chapter.

How bigger global scope can be acquired through composing process depends mainly on the characteristics of the base layer AUs. As some good conditions, it can prevent middle level composite from dropping into a trap of local optimal structure. So, the composite model with its global scope could be useful for AI design and solving problems of machine learning, because global scope is seeking for global divergence in this area. Under some good conditions, there can arise the middle level autonomous units on the stratification. It could be a good representation of geometric structure of organization, in particular social organization.

6.2.9 Higher Stratification and Its Number of Levels

Combining the composite frame and the local rule of Decentralized Autonomous Units, as makes composing process repeatedly activate, the structure of composite is stratified in each plane and layer. So, there appear hierarchical structure and geometric characteristics. This hierarchical structure is not designed but spontaneous stratification. From the results of pilot simulation, the height of the stratification is limited under the condition of finite number of units. And, it looks unstable, which depends, however, mainly on the initial conditions and continuous action of searching and evaluation.

Figure 6.5 is re-illustrated for this chapter, using same video source as the video presentation in Kashiwabara et al. (1993), and the figures 3 and 4 in Yamamoto (1994b).

In the Fig. 6.5, it expresses the stratification of composite from the autonomous units illustratively. The implemented simulation was running by the original program written by C language (Kashiwabara et al., 1993), and the colored stars and connecting lines showed the stratification and generating higher ranked composite [5]. In this illustrated figure, AUs are searching good partners and evaluating them, then connecting each other. At that time, the first layer composite (level 1-th) arises on the space. Composite 1-th takes same behaviors and connects to each other at the first layer. For instance, composite 2-th arises on the second layer. THE contained AU and composite 1-th are continuously evaluating the vertical connectivity and recomposing the composition.

Here is only one composite 4-th in the figure. In fact, in running of the simulation, it is so infrequent that there arises composite 5-th and over composite 6-th without suitable parameter tuning for higher stratification. This phenomenon on the display shows that natural stratification has some limitation of hierarchy and it is the optimal stratification, and that there occur local trap of optimization at the third layer or fourth layer.

6.2.10 Behavior of Composite

Formally building and logically driving the composite model, we can observe the behaviors of hierarchical system. The height of stratification changes and fluctuates vertically, in accordance with composition and decomposition. It is accompanied by horizontal dynamics of DCAS. It is also our trial to investigate the relation between global characteristics of higher composite and local characteristics of lower composite.

From the local rule setting, each endowment is variable according to change of input. And, from the open condition of the frame, if another AU (AU2) is encountering to the AU (AU1) independently, even though the AU (AU1) has evaluated all values and connected to the other AU, the AU (AU1) should recalculate

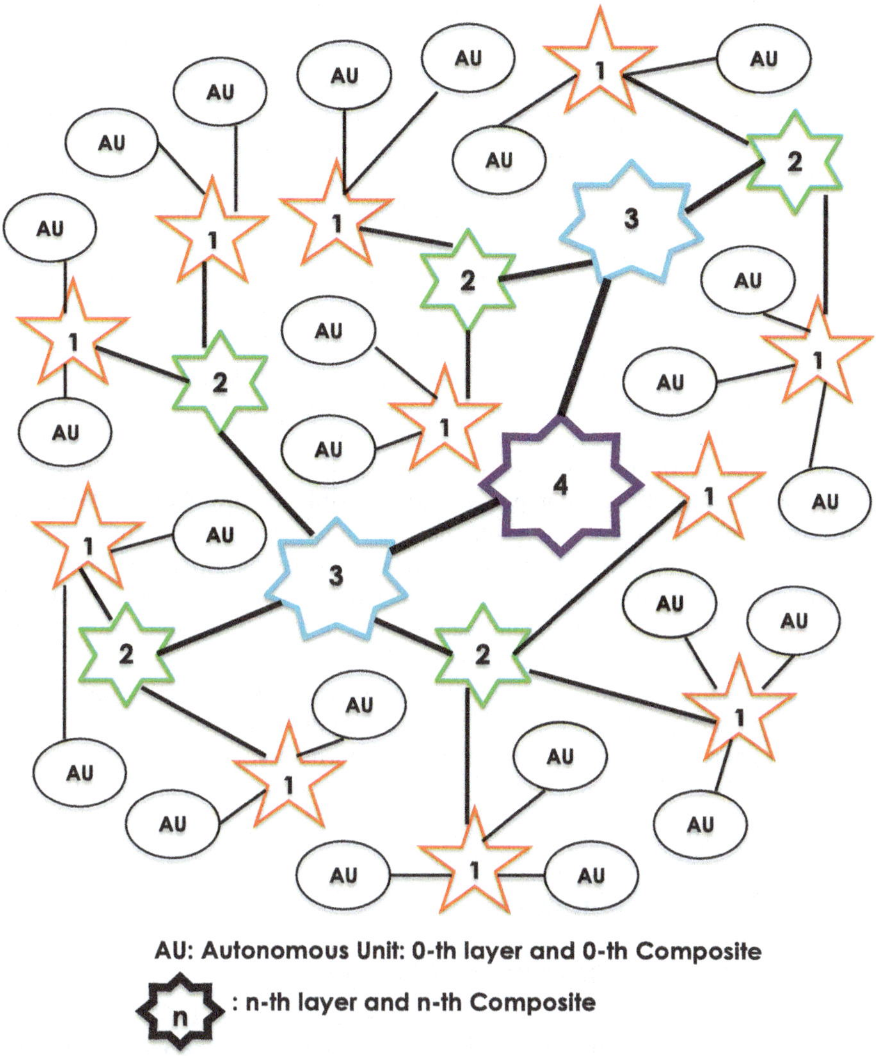

Fig. 6.5 Re-illustrated image of the simulation of stratification

and evaluate every AU in the scope. Adding to these changes, from the nest condition, recursive and repeated operations in composing process makes each operator of composites change.

Surely, the l-level m-th composite (Clm) should re-evaluate every composite vertically and horizontally, and re-connect to new composite, and re-perform the task in accordance with the same situation changing. Thus, computational complexity is supposed to become bigger rapidly in this calculation. And, the simulation shows continuous transformation of composite.

6.2.11 Optimal Structure in Stratification

Fluctuation of stratification shows optimal process, based on active move of composition and decomposition. It is, however, unstable in the simulation, because continuous acting of AUs brings rapid rotation of composites from arising to annihilation.

However, spontaneous organizing simultaneously with information processing makes an optimal organization for task processing. Processing in concurrence with composing is efficient, and forms optimal composite. Composing process gives efficiency of processing and optimal performing of task in this meaning.

So, composite is one of systems models of dynamics by information (Yamamoto et al., 1992). In organization, we shall see neither stable nor optimal structure, but continuously optimal transformation of organizational structure.

Transformation of composite itself consists of optimal process. When we pay an attention to the top of composite in stratification, it presents optimal transformation. If we identify the top of composite with Possible Future World (PFW), the transformation of the top of composite shows bifurcation to possible future.

A completed composite has global and higher dimensional structure. The outlook is hierarchy and inside is nested. Each composite has same structure, horizontally and vertically. It seems to be fractal geometry. The geometric structure generates functional characteristics and phenomena.

Through composing process, geometric structure of composite is continuously transformed from "distributed" to "hierarchical." In this situation, some measurable quantity of geometric structure varies as the transformation. At first, the measurable quantity will determine own form of geometric structure, when it is generated.

6.3 A Model of Complexity

6.3.1 Geometry of Structure and Complexity

A system has own geometric structure which represents its characteristics. It is not always stable. We should emphasize the optimal transformation of structures rather than structural stability. Transforming of systems shows a variety of geometric structures. Hierarchical structure is one phase that produces efficiency and complexity at the same time. Distributed structure can produce same features by another way. It depends on connectivity and network topology.

Bifurcation has own geometric structure and numeric characteristics on the tree structure. Each bifurcation point and its neighbors can show local structure of composite and future information. The whole of the trees shows global structure. Many bifurcations and branches cause the complexity by themselves. While number of branches is increasing, it must require a lot of computation time for predicting.

Thus, it makes the complexity grow up. Sometimes, the tree structure obeys features of the graph theory.

When we can observe any geometric structures such as fractal, foliation, and manifolds in systems structures and processes, we can enjoy the characteristics of these advanced geometries. On the other way, if we can describe them as a discrete form, we will enjoy the fruitful results of discrete mathematics such as cellular automata.

6.3.2 Hierarchical Structure Breaking

When hierarchical system or structure consists of remarkable layers and boundaries, hierarchy breaking is a cause of complexity as well as symmetry breaking. This hierarchy breaking occurs over plural hierarchical layers. The relation across the hierarchical layers may produce complexity.

These complex phenomena through the stratification often appear on many real fields, such as biological systems, a human body itself, inside of computer, working network systems, social systems and movement of air from the atom to the stratosphere.

6.3.3 Consideration of Complexity

With the progress of systems theory, researchers have paid their attentions to complex systems. In particular, it is expected to obtain some emergent characteristics. In this research, we consider two kinds of complexity. First, it is structural complexity. Geometric structure shows complexity. Hierarchy has some complexity. Decentralized and distributed structure has also complexity. Some process on the structure of fractal and foliation are deeply connected to complexity and chaos.

Adding to structural complexity, process and operation often generate complex phenomena. Repeating and recursive operation can make complex process. Second, when we try to make a computational model of the structures and processes and realize them on a computer display, we will immediately face to computational complexity. It is calculation complexity.

On the other hand, many systems for information processing like a hierarchical system are usually planned to reduce the complexity. So, there is a paradoxical problem. Also, when we seek for efficiency and some exclusive advantage in computation space, there shall arise imperfect information and uncertainty to the future. If we try to erase this uncertainty, it will change to complexity in the processing process.

We can easily find complex systems in the natural world. Bio-system is a typical complex system. In human body, the immune system, the nerve system and neural network, the metabolism system, and the endocrine system, all of the bio-systems are recognized as a complex system. So, we can describe the functional processes and the action mechanism of local part of the systems as some relatively simple causality relations and build operational models. However, the four systems in human body are so interactive that we cannot image the global mechanisms and processes of appearing of some symptoms. It is one of the complexity of bio-systems.

6.3.4 Composite and Complexity

According to the Chap. 2, the concept of composite contains two important conditions, nested stratification and repeating and recursive operation of composing. Both conditions will form and show some kind of complex characteristics. The nested and repeated recursive operation immediately relates to dense superposition structure which may generate complexity.

Bifurcation is one of the representations of complexity. It simply shows that some entity has plural future images. It is usually not alternatives to choose. No one can choose a branch of the bifurcation. It should be inevitably proceeding to the future without choosing. So, it always contains uncertainty. The number of branches shows complexity. While the number is increasing, the complexity is growing up. To what extent it grows depends on total number of the branches in bifurcation. Although we are unable to choose one branch, we are able to know the possibility where to go and what to happen in near future.

Thus, it is subjective images of future and subjective future itself. Bifurcation of composite presents its subjective future to us. It is an image of transformation of the composite. Figure 6.6 expresses the complexity of bifurcation. Many branches represent the complexity to the future in this bifurcation model. So, it is complex future. If each bifurcation point relates to composite, the branches show the future possibilities of each composite. Also, transition of composite is predictable transformation of composite in near future.

Figure 6.6 is originally illustrated for this chapter.

6.3.5 Generation from Complex and Chaos

As well as many information processing systems, designed hierarchical systems are basically expected to be a device to reduce complexity. On the other hand, spontaneous hierarchical systems, in particular stratified systems can contain complex layers. In this case, hierarchical systems themselves are considered as a complex system. Surely, even designed or artificial hierarchical systems transmute to complex system, according to complex structure formed by flow of processing.

Three dimension of a composite, vertical, horizontal virtual and nested stratification, contribute to forming complexity. Also, it is the source of complexity to repeat

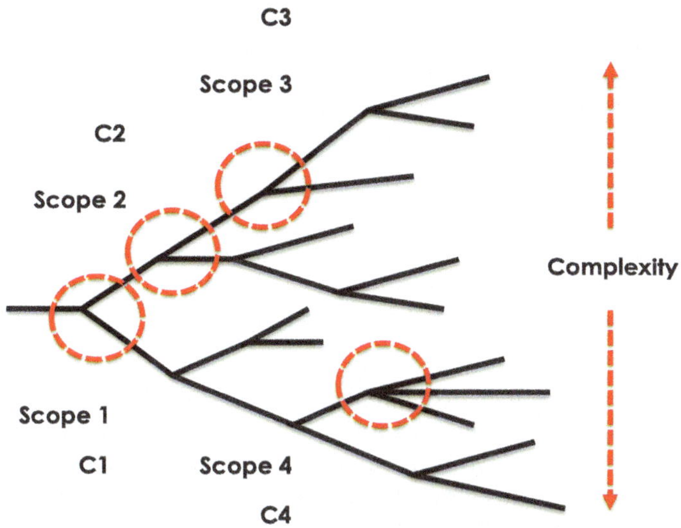

Fig. 6.6 Bifurcation and composite

simple operations and vertical recursive operations based on horizontal DCAS and network structure. So, complexity is mainly caused by both spontaneously generated structure and the process which contains sub-processes of recursive operator and repeat operator. Designed Hierarchy has both structural complexity and process complexity of flow. Spontaneous stratification has process complexity of composing.

6.3.6 Paradox of Complexity

Generally, systems performing should reduce complexity through the information processing, whether it is a human organization or computer system. In particular, a hierarchical system is designed as can process complex information. Through the hierarchical layers vertically, it will be filtered, condensed, and summarized to create alternatives. Also, stratification based on decentralized autonomous units keeps so wide and flexible diversification that it can process pluralism problems with bad structures.

Nevertheless, some system generates complexity, although the system has the ability to reduce the complexity. Systems complexity and organizational complexity are generated from the geometry of structure and information flow in the process of performance. Thus, complex system is dealing with a task, processing information, while it is generating complexity.

Complex phenomena might be generated from some process and flow of processing. System designers and AI designers are at times prone to evaluate hard architectures as higher than they really are. Flow of processing and its locus makes system structure and produce complexity. In both designed hierarchical systems and spontaneous stratifications there occur complex layers.

6.4 A Model of Prediction

6.4.1 Imperfect Information and Prediction

Through human history, imperfect information processing needs oracle and prediction from the maidens of Athens to Turing machine. So, we should prepare a model of prediction which is able to be used in the computation. Increasing complexity erases the possibility to go along a linier transit to the future. However, increasing number of branches of bifurcation enables us to see the another future world. In the area of future forecasting, there are many ideas such as stochastic model, big data analysis, and ML type AI. A data-driven model is a main stream. It is necessary for this way to use probability in any subprocess of prediction.

Big data analysis treats all events as if they dealt with together, even though each event has each cause of occurring. In some sense, it is a black box. Classic view of science requires each mechanism and rational induction process to the reason of occurrence. It is not necessary for contemporary systems science. Without stochastic information and probabilistic data, can we add absolutely new information to the prediction system? It is our purpose of the research of prediction.

6.4.2 Prediction and Bifurcation

If we can perceive the future as bifurcation, the prediction problem is to be discrete and finite. However, we cannot choose one by our intention from the branches. Bifurcation is the way of limiting possible future to a few branches. So, it is, as it were, possible but passive future. This future prediction model is not very simple.

First, we can predict possible future in the branches of bifurcation. It means that all possible futures consist in the branches at that point. Second, surely from present to future, process is going along the branches. Third, we are unable to choose one in the branches. We have to accept the result. Using a stochastic model, we can calculate the possibility at each bifurcation point.

As for imperfect information processing, generating a branch is important rather than cutting a branch to which AI (Artificial Intelligence) and any programs always pay attention. The way of generating a branch has two types. One is generating a new branch from genuine complexity. The other is artificial setting a generator. ML

(Machine Learning) type AI needs some generator and generating new information. In fact, stochastic type generator is often introduced to the AI. Thus, often ML type of AI implicitly contains stochastic characteristics. It is unchangeable that we need absolute generation of information.

6.4.3 Bifurcation and Its Characteristics

At each bifurcation point, there occur such the operations as diverging branches, crossing, annihilating, and conjunction. Cutting a branch is the way to reduce uncertainty of prediction process. So, on the process of prediction, such as AI, cutting a branch is effective way to reduce both uncertainty and computation time. Diverging a branch means growing up of uncertainty, simply as generates complexity. Annihilation is also simply reducing uncertainty (See Fig. 6.7), but means disappearance of one possible future. Crossing and conjunction show complexity. The two operations are not assumed at usual tree structure in Graph theory.

Figure 6.7 describes bifurcation and its operation. Diverging a branch shows increasing some complexity. On AI and decision-making theory, cutting branches of the tree should be always good methodology for problem solving. On the contrary, creating a branch or spontaneously generating a branch on the bifurcation or the decision tree is now the proper path to reach to global solution or better solution.

Figure 6.7 is originally illustrated for this chapter.

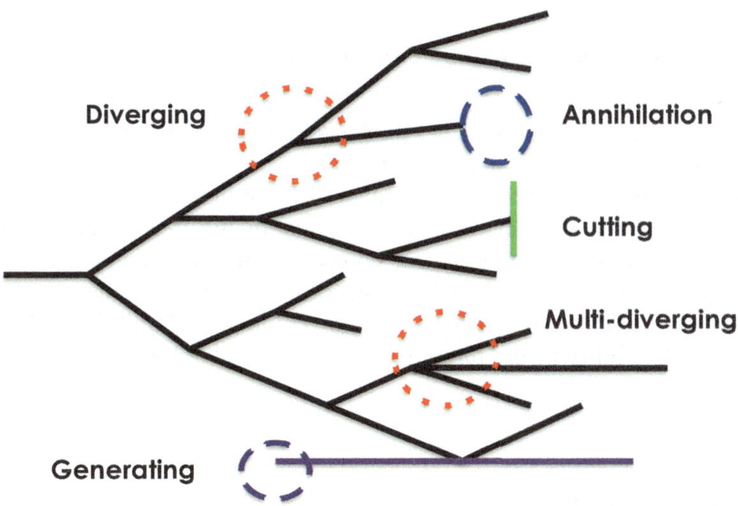

Fig. 6.7 Bifurcation and its operations

6.4.4 Bifurcation and Composite

Basically, the composite model has a meaning of overcoming imperfect information. In the composite model, each autonomous unit takes a behavior of gathering necessary resources and information for problem solving. Through the moves of the autonomous units, there naturally appear the wider base of decentralized autonomous units and higher stratification of composite. Stratification brings us more precious and possible prediction. Apparently, aggregation (union) of local scopes can get necessary information over a scope of the individual autonomous unit. So, wider connection of composites brings global scope. Higher stratification generates the wider scope.

When bifurcation means the possible trajectory of a composite, each bifurcation point is related to each future of a composite. Also, the global scope of a composite covers its bifurcation point. The global and higher scope of a higher composite can see a farther future point. When a branch of the bifurcation shows a possible future world (PFW), we have the two models. First, each PFW shows a phase of transformation of one composite. Second, each PFW is not related, because each PFW is generated by an invisible generator genuinely. In appearance, PFW looks transformed continuously. However, they are independent.

Surely, on computer simulation, number of branches is increasing, as must require a lot of computation time for predicting. On the other side, it means that future possibility is growing. We can profoundly comprehend the whole system of possible future worlds.

6.4.5 Bifurcation on Top of Composite

What is the bifurcation and branches on the composite model? Diverging branches show prediction of transformation of the composite. When the transformation is stochastic, the branches mean probabilities of the transformation and next future of composite. So, going forward to next point in the tree means nothing less than transforming the composite and composing the new composite. The past trajectory of bifurcation points shows transformation of the composite. The other points and their branches mean possible future composite. Plural bifurcation means complexity. Possibility of the future sets the limit to plurality. So, this complexity has some limitation.

6.4.6 Bifurcation and Complexity

Diverging branches on bifurcation can be defined as one of the complexity (See Figs. 6.6 and 6.7). Adding to it, the topology of bifurcation makes some complexity, such as loop, recursive, crossing, and joint. It is surely converse to cut the branches on bifurcation, as is usual methodology of AI. In fact, AI technology needs both ideas. One is to diverge branches. The other is cutting the branches. In this meaning, complexity is deeply related to intelligence and AI.

Growing bifurcation, diverging a branch, and the branch prediction makes a set of future prediction methodology. Future prediction needs creating information for an image of the future. It means that concept of future occurs, imperfect information problem happens.

On the other words, bifurcation is a fine and sophisticated model of future. It is an artificial production of future prediction. Plural branches show low-level complexity. Unexpected branch diverging is high-level complexity. On the graph of bifurcation and branches, if some global scope can cover any branches which may yield at the bifurcation point, it is called as predictable.

6.4.7 *Uncertainty in Composite*

Each entity (AU) has own limited scope for searching necessary information (See Fig. 6.1). So, the first uncertainty of information is caused by "out of scope." It depends of the rage of a scope. And, it leads AUs to make composite. The second uncertainty is future uncertainty. While time is going forward, uncertainty to tomorrow is arising in information space. If we do not care what will happen to us tomorrow, uncertainty will disappear from our recognition. In a steady state world, there is no concept of time, no future, and no uncertainty.

The third uncertainty is original in a system. It depends on the characteristics of structure and process of a system. Some structure and geometry may generate complexity and chaos. They easily become uncertainty. On the contrary, complexity is expected to generate enough good information to compensate for uncertainty and imperfect information. Hierarchical system has two aspects. One is to absorb complexity an uncertainty of information. The other is to generate structural complexity and inherent uncertainty. On composite model, each AU takes a behavior of connecting to resolve imperfect information problems, even though range of the scope is limited. Composing process makes some complexity and chaotic phenomena by the acts of repeating operator and recursive operator.

6.4.8 *Complexity and Self-Organization*

After the concept of self-organization, system theory has sought to new ideas. Complexity and chaos is a couple of the strong candidates. We gave attention not only to the profoundly interesting phenomena and strange behaviors of complex system and chaos, but also to the possibility to generate newly genuine matters.

As system recognition, self-organization is an autonomous system which can organize itself without any control and design. As informational recognition, self-organization is that it has no information for organizing and forming order from chaos. And, self-organizing reduces entropy against the law of increasing entropy, as is making order from disorder or chaos. So, we expected that an abstract concept of self-organizing system could provide creative information and newly genuine matters to our information processing systems.

When we see the result of self-organization, we can recognize some meaningful pattern or useful structure of geometry or topology. Cybernetics as system theory has also self-control function. Structural dynamics is supposed to generate own characteristics. It depends on structural stability. Transformation of system often generates optimal structure. Transformation itself has the function of self-generating.

6.4.9 *Artificial Intelligence and Imperfect Information*

Artificial Intelligence is at the zenith of astonishing prosperity. What is more astonishing is that AI is made up of classic models and implementation technology, such guiding principles as Perceptron, Neocognitron, neural network, Boltzman machine and Kullback–Leibler divergence, and methodology of fast differentiation. Unknown information can be obtained by image processing, using AI. However it may compute and process the program well, it needs generating information (See Fig. 6.7).

Usually this type of AI uses stochastic methodology and computational process which implicitly contains probability. In other words, AI needs generating information at somewhere in the computing process. For example, GANs (Generative Adversarial Networks), is now a powerful tool of algorithm for Unsupervised Learning of ML (Machine Learning). GANs have some generator. Transcendental Genesis is the artificial conjecture for problem solving of imperfect information such as future prediction, creating oracle, and bifurcation forecasting.

6.5 Conclusion and Developments

Using a composite model and its frame, I will try to investigate next problems. First, I will prepare a variety of local rule sets and simulate the composite model with new local rule set, using high performance computer. It is expected that contrasting type of geometry appears on the simulation. Classification of geometry of structure will be fruitfully proceeding.

Second, I will investigate the properties of continuous model of composite, and to the behaviors and patterns of discrete model of composite. Properties of transformation of continuous composite are to be precisely investigated. Third, visualization of behaviors and geometric characteristics of composite is very interesting and useful for understanding of complexity and information processing. At last, I will try to build an operational model of prediction based on decentralized autonomous units. Grounded on these researches, conceptual, mathematical, and philosophical investigation into Transcendental Genesis and Possible Future World is the last aim of this research.

Acknowledgments I would like to express my heartfelt thanks to late Prof. Takehiko Matsuda, Prof. Yasuhiko Takahara, Prof. Bunpei Nakano, and Prof. Junichi Iijima.

System finding is the first step of system recognition. As Norbert Wiener did, we shall try to find it and describe general formation for anonymous black box. As Herbert Simon did, we shall set it up on the fruitful conceptual world, named as artificial sciences. As prof. Matsuda did, we will apply it to organization and management, and build a new concept of intelligence such as organizational intelligence. Also, prof. Takahara did, we must comprehend the whole system by abstract representation and mathematical operation. The logical developments and scientific trials of systems science are continuing to the future, even though it will change the name by itself. Again, let's find it out from invisible space. A new system model implies in regular phenomena and irregular phenomena, also in, prima facie, empty space.

References

Kashiwabara, K., Yamamoto, T., & Nakano, B. (1993). An algorithm for generating artificial society and the simulation model of artificial creatures society. In *Proceedings: The 4th Autonomous Distributed System Symposium* (pp. 125–130).

Simon, H. A. (1999). *The sciences of the artificial* (3rd ed.). The MIT Press.

Wiener, N. (1948/1999). *Cybernetics: Or control and communication in the animal and the machine*. The MIT Press.

Yamamoto, T. (1994a). *A study on complex distributed composite and self-generating — Process of complexity and emergence on social phenomena and living phenomena.* Doctoral Thesis, Tokyo Institute of Technology, Oookayama, Tokyo.

Yamamoto, T. (1994b). Artificial society: Atrial to generate society artificially and virtually. *Sociological Theory and Methods, 9*(2), 203–218.

Yamamoto, T., Nakano, B., & Matsuda, T. (1992). System, information and self-dynamics. In *International conference on economics/management and information technology (CEMIT92/CECOIA3), proceedings* (pp. 211–214).

Part II
Systems Management

Chapter 7
What Should Be Added to Science for Solving Wicked Problems?

Junichi Iijima

Abstract In today's uncertain and unpredictable society, known as the VUCA era, the world is facing at the frequent occurrence of severe disasters caused by huge typhoons and torrential rains due to global warming, the global COVID-19 pandemic, the UK's withdrawal from the EU, and the trade friction and struggle for supremacy between the US and China, in addition to such large-scale complex global problems, we are surrounded by so-called "wicked problems" that are difficult to solve, for which there may or may not be a single correct answer.

To what extent is science effective in solving these "wicked problems"? What can complement traditional science, which has focused mainly on "nature", to solve the "wicked problems" facing humanity? This is the question we will consider in this essay.

I believe that we can solve the "wicked problems" by complementing the "science" base with an empathy-based approach from the perspective of "design".

Our approach starts from the "wicked problems" of people, organisations and societies, clarify the current situation based on empathy, conceive an artificial system to support change it into a desirable future, create a blueprint, implement it in society in cooperation with engineering and administration, and then make the solution sustainable.

This approach is a combination of a general systems development approach and design thinking. In this chapter, we discuss the background to this approach.

First, I will summarise what the concept of "wicked problems" is, then I will clarify the author's position on "science", and then I will trace the evolution of systems theory as a complement to it. Finally, the importance of empathy as an "alpha" that should be added to science for solving wicked problems is discussed, after showing how it is linked to "design".

J. Iijima (✉)
Department of International Digital and Design Management, School of Management, Tokyo University of Science, Tokyo, Japan
e-mail: junichi.iijima@rs.tus.ac.jp

K. Kijima et al. (eds.), *Systems Research II*, Translational Systems Sciences 27,
https://doi.org/10.1007/978-981-16-9941-2_7

Keywords Wicked problems · Science · Structural similarity · Soft systems methodology · Design thinking

7.1 Introduction

Since the beginning of the 2010s, Japan has been hit by a number of major natural disasters, including the tsunami caused by the Great East Japan Earthquake, the explosion of the Fukushima nuclear power plant in the Tohoku region, and huge typhoons, torrential rains and heat waves caused by global warming. These natural disasters are not necessarily confined to Japan, but occur in many other parts of the world, such as the massive forest fires in southeast Australia at the end of 2019 and the floods in Germany and Belgium in July 2021 and the global pandemic of COVID-19, which began in 2020. On the international front, there is Brexit in the EU, trade friction and the struggle for supremacy between the US and China, the withdrawal of US troops from Afghanistan at the end of America's "longest war" starting with the attacks on the US, and the diversification of values through the use of the internet and social media.

In addition to these massive complex problems, we are surrounded by so-called "wicked problems" which are difficult to solve, for which there may or may not be a right answer, and for which there is no single solution. Thus, today we live in a highly uncertain society, called the age of VUCA (Volatility, Uncertainty, Complexity, Ambiguity). How effective is science in solving these "wicked problems"? What can complement traditional science, which has focused primarily on "nature", to solve the "wicked problems" facing humanity? This is the question we consider here. We believe that the concept of "design" gives us an important clue to this. We are trying to solve "wicked problems" by taking "science" as a base, and complementing it with an approach based on empathy, from the point of view of "design".

There are many definitions of the word "design". One of the best-known definitions is that of "the intellectual activity of creating material artefacts", as proposed by Herbert A. Simon (Simon, 1996), who was awarded the Nobel Prize in Economics in 1978. Here, I would like to extend this a little and regard it as "the intellectual activity of creating artificial systems".

The author entered the Tokyo Institute of Technology in 1973 and joined Professor Takahara's laboratory in April 1976. Prof. Takahara was an assistant professor at the time, and the chair was held by Prof. Takehiko Matsuda, who later became President of the Tokyo Institute of Technology. It was Professor Herbert A. Simon who was Professor Matsuda's supervisor at Carnegie Mellon University.

As shown in Fig. 7.1, our approach for solving "wicked problems" starts from the a "wicked problem" that people, organisations, and society has, identifies the problem, and creates a "blueprint" for an artificial system (material products, services, social systems, etc.) that can solve them. We then work with engineers and governments to implement these solutions in society and make it sustainable.

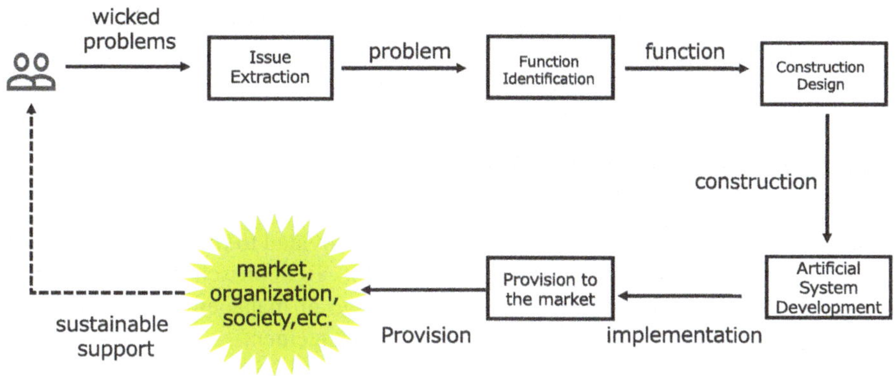

Fig. 7.1 Approach for solving "wicked problems"

7.1.1 Issue Extraction

This is the phase in which issues are identified from the problematic situation based on empathy. In a global society, the issues to be addressed are not limited to a single region, and consideration for diversity and cultural understanding are essential. It also requires the "insight" into the subject matter that is required in "design".

7.1.2 Function Identification

This is the phase where, based on the "empathy" from the extracted problem, we identify the functions that an artificial system should have as a solution.

7.1.3 Construction Design

This is the phase in which the construction of an artificial system is logically designed to realize it from identified functions. In particular, it is very important to use digital technology to solve the social problems of modern society.

7.1.4 Artificial System Development

This is the stage in which, based on the blueprint of the designed artificial system construction, we develop it in cooperation with engineers and administrative bodies.

An artificial system is a material product, a service, a social system or any other artificial system.

7.1.5 Provision to the Market or Society

This is the phase in which the artificial system is provided to the market or society so that users with problematic situations can use it. To approach the process up to this point, it is essential to have a basic knowledge of management, including an understanding of the market and organizational management.

In this essay, I will first summarise what the "wicked problem" is, then clarify the author's position on "science", and then explain general systems theory, which emerged as a complement to science. I believe that general systems theory is composed of two pillars: "emergent properties" and "structural similarity". By briefly explaining these two pillars, I will clarify the characteristics of general systems theory.

Since the 1960s, there has been a movement to apply general systems theory and systems engineering for solving social problems. This is called hard systems methodology. Unfortunately, it was not successful, and later the approach of applying the ideas of general systems theory and so-called second-order theory for solving social problems developed into soft systems methodology, engineering systems, and design thinking.

In this essay, I conclude that "wicked problems" can be solved by complementing "science" with an approach based on "empathy" from the perspective of "design".

7.2 What Are Wicked Problems?

First, let us consider what we mean by a "wicked problem". The term "wicked problem" emerged from the field of policymaking in the late 1960s to describe a complex social problem that is difficult to solve and for which there may or may not be a single right answer.

In the preface to Management Science in 1967, Charles W. Churchman quoted Horst Rittel as saying at a seminar that "the term "wicked problem" refer to that class of social problems which are ill-defined, where there are many clients and decision makers with conflicting values, and where the ramification is the whole system are thoroughly confusing. The adjunctive "wicked" is supposed to describe the mischievous and even evil quality of these problems, where proposed "solutions" often turn out to be worse than the symptoms" (Churchman, 1967).

After that Horst W. J. Rittel et al. listed up ten distinguishing properties of wicked problems in "Dilemmas in a General Theory of Planning" (Rittel & Webber, 1973).

1. There is no definitive formulation of a wicked problem.
2. Wicked problems have no stopping rule.
3. Solutions to wicked problems are not true-or-false, but good-or-bad.
4. There is no immediate and no ultimate test of a solution to a wicked problem.
5. Every solution to a wicked problem is a "one-shot operation"; because there is no opportunity to learn by trial-and-error, every attempt counts significantly.
6. Wicked problems do not have an enumerable (or an exhaustively describable) set of potential solutions, nor is there a well-described set of permissible operations that may be incorporated into the plan.
7. Every wicked problem is essentially unique.
8. Every wicked problem can be considered to be a symptom of another problem.
9. The existence of a discrepancy representing a wicked problem can be explained in numerous ways. The choice of explanation determines the nature of the problem's resolution.
10. The planner has no right to be wrong.

In the 1990s, the term "wicked problem" came to be used not only in the field of policymaking, but also in design studies and corporate strategy. Buchanan, for example, refers to design problems as a wicked problem "because design has no special subject matter of its own apart from what a designer conceives it to be" (Buchanan, 1992).

In addition, Camilus (2008) states that Walmart's problems are wicked and there is no systematic management for them because of the characteristics such as the existence of "many stakeholders with different values and priorities" and "the challenge has no precedent" (Camillus, 2008).

What they all have in common is how to bridge the barriers between diverse values and opinions.

Cocklin states that developing a common understanding and consensus on an issue is the way to deal with a wicked problem. What is required here is a dialogue between the people involved and a story that people can feel has a consistent meaning (Conklin, 2009).

7.3 Can Science Be Directly Applicable to Solve Wicked Problems?

7.3.1 Science

Let me summarise my position on how I see "science".

There have been various debates in the field of philosophy of science about what science is, but Checkland summarises the classical cycle of positive hypothesis-testing research in natural science as shown in Fig. 7.2 (Checkland, 1991). Since it is circular, let us start from researcher's attempt. The researcher attempts to test a hypothesis about the area of concern A, and conducts an experiment or other verification attempt in a field related to A, which is a selected part of an external reality outside himself/herself. The results of the observation are repeatable ones. Some of the results may be refuted as invalid, or they may be considered intact. In this way, new hypotheses are generated and new knowledge about A is accumulated. This is the argument of Fig. 7.2.

Furthermore, Checkland mentions the three Rs, Reductionism, Repeatability, and Refutation, as the scientific views that support this approach. Reductionism has three meanings: (1) reductionism in experiment, (2) reductionism in explanation, and (3) reductionism in thought.

1. Reductionism in experiment is the idea that experiments should be conducted in a completely controlled environment, not in the complex and diverse environment of the real world.
2. Reductionism in explanation, represented by the so-called "Ockham's Razor", means that "the simpler the explanation, the better". It also includes the idea that all explanations should be given in the terms of physics, the king of all sciences.

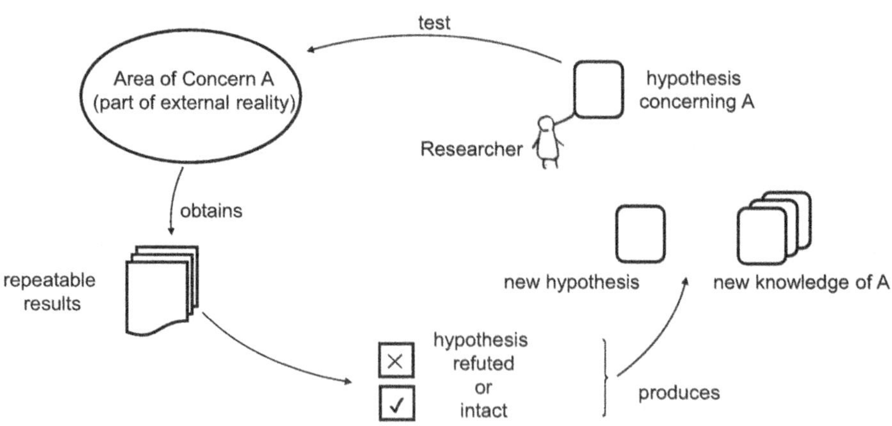

Fig. 7.2 The cycle of positivist hypothesis-testing research in natural science

3. Reductionism in thinking is elemental reductionism, which divides an object into elements and says that the whole can be understood on the basis of understanding of each element (Checkland, 1981).

Repeatability means that only what arises from experimentation can be accepted as "public knowledge", and therefore presupposes neutral observations that can be shared by all.

Finally, refutation means that whether or not a statement is scientific depends on whether or not it can be refuted, and only a statement that can be "refuted" is a scientific statement. For example, the fortune-telling statement "If you are born under the sign of Virgo, your lucky color today is yellow" is not a scientific statement, because the interpretation of the meaning of "lucky" is ambiguous and cannot be refuted. On the other hand, the statement "the sun rises in the east" can be refuted if it rises from a place other than the east, so it can be considered a scientific statement, where it is noted that the author is not claiming that fortune-telling is meaningless.

Checkland sums up scientific activity as "we may *reduce* the complexity of the variety of the real world in experiments whose results are validated by their *repeatability*, and we may build knowledge by the *refutation* of hypotheses" (Checkland, 1981).

Incidentally, the debate on "what is science?" began in the twentieth century with the development of logical positivism by the Vienna Circle, which culminated in what is called the Received View (RV) (Suppe, 1977).

RV argued that scientific theory is axiomatised in mathematical logic, where it consists of theoretical terms, observational terms, a corresponding rule between them and logico-mathematical terms.

RV was later challenged by a group of so-called "worldview perspectives", such as Paul Karl Feyerabend, who argued that science is "anything goes"; Norwood Russel Hanson, who denied the existence of neutral observation, arguing that "observation is theory-laden"; and Thomas Samuel Kuhn, who boomed with his "paradigm theory". As a result, it lost its momentum. And in the field of philosophy of science, the interest shifted from the debate on "what is a scientific theory" to the social behaviour of the group of scientists who produce the scientific theory.

7.3.2 Limitation When We Apply Science for Wicked Problems

Checkland also discusses the possibility of applying the scientific view represented by the three R's to a system that targets human activities when trying to solve "wicked problems" (Checkland, 1981).

Let us start with Reductionism. Just as, for example, the "climate" of different teams and organisations cannot be derived from the properties of their members, so we cannot understand a whole based on the properties of its constituent elements alone, but must consider the emergent properties of the interactions among them.

It is necessary to consider the emergent properties that arise from the interaction between elements. For this reason, it is difficult to apply Reductionism to systems that are subject to human activities.

Next, let us consider repeatability. In the case of a system including human activities, the presence of an observer influences the behaviour of the observed object, which in turn changes the results of the observation. Therefore, it is unreasonable to assume repeatability in the case of systems including human activities.

Then we should not discover rules and laws from phenomena, but we should take the structure of the phenomena as a problem, and this is the reason why the systems theory with emergent properties and structural isomorphism as its two pillars was born.

Finally, what about Refutation? This involves the problem of the interpretation of concepts in discourse. There is no shortage of examples of terms whose interpretation is itself controversial in the discourse on systems, including human activities. Given this, there may be problems in applying it to Refutation as well.

7.4 General Systems Theory and Emergent Property

Thus, as a complement to "science", systems theory was born with the concept of emergent properties and that of structural similarity as its pillars. Checkland states that systems thinking consists of a combination of two concepts: the concepts of Emergence and Hierarchy, and the concepts of Communication and Control (Checkland, 1981) Here, the author will focus on three topics, Emergent property, Cybernetics, and Structural similarity, and explain them briefly. The concept of Emergent property is described by Bertalanffy in his book *General System Theory* (von Bertalanffy, 1968).

In contrast to the traditional reductionist approach, such as that of physics, the biologist Bertalanffy took the organismic position that the whole is more than the sum of its parts, and envisaged a unity of the sciences, focusing on emergent properties, but also on structural similarity.

The Society for General Systems Research was founded in 1954 by Bertalanffy, the biologist Anatol Rapoport, and the economist Kenneth E. Boulding, who was later awarded the Nobel Prize in Economics. In its prospectus, the Society stated the following four objectives of general systems theory as shown in Fig. 7.3:

1. to study the isomorphism of concepts, laws and models, and to spread them effectively to other fields,
2. to encourage the development of theoretical models in areas where they are lacking,
3. to avoid duplication of theoretical research efforts,
4. to unify the sciences by improving communication between experts.

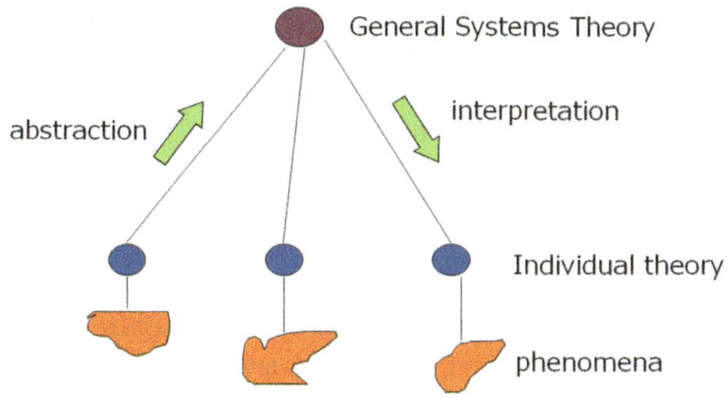

Fig. 7.3 Way of Thinking in General Systems Theory

You can see from this that the concept of structural similarity is a central topic in systems theory. This attention to structure is, so to speak, a characteristic spirit of the twentieth century.

7.5 Cybernetics

The concept of "cybernetics" was first introduced to the world in the book *Cybernetics: Communication and Control in the Animal and the Machine*, first published in 1948 (Wiener, 1948). In the preface to the book, he says that he coined the neologism from Greek as a name for the field because he believed that a series of problems revolving around communication, control and statistical mechanics, whether in machines or in living tissue, can be essentially unified.

The origin of the name "cybernetics" has already been mentioned, but the concept can be briefly summarised as follows: the coordination mechanism by feedback of information, as shown in Fig. 7.4, can be found in many parts of the world, including animals and machines.

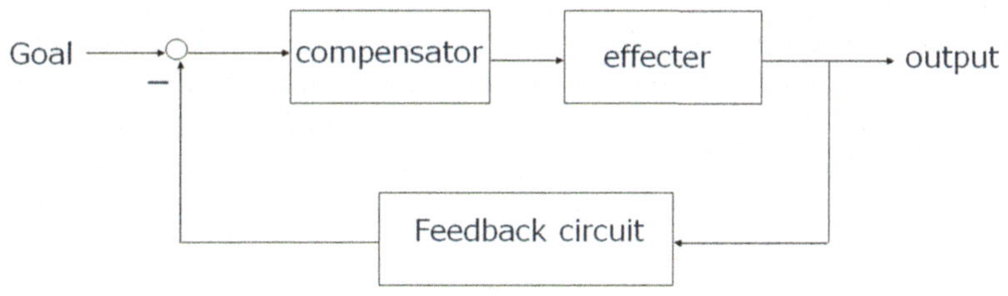

Fig. 7.4 Negative feedback

The word "cybernetics (cybernetique)" is said to have been first used by the French physicist André-Marie Ampère in his essay of *Essai sur la philosophie des sciences* in (Ampère, 1834). It is said that the term was used there to refer to the means of governance in political science.

According to Nobert Weiner, the word "cybernetics" comes from the Greek word "κυβερνήτης", meaning "governer". The reason for choosing this term is said that the first paper written on feedback mechanisms was Maxwell's work on governors, and that the word governor comes from the Latin accent of κυβερνήτης.

As already introduced, "cybernetics" simply means "control by feedback, which is commonly found in animals and machines", and feedback here means negative feedback as shown in Fig. 7.4. Simply speaking, negative feedback is that the input is adjusted so that the deviation of the output from the goal state is reduced. An example of this is the physiological regulation of body temperature and blood pressure, known as homeostasis.

Thus, when we speak of cybernetics, we usually mean negative feedback, but there is also a concept of positive feedback. This is known as second cybernetics, a concept introduced by Magoroh Maruyama in "The second cybernetics: Deviation-amplifying mutual causal processes" (Maruyama, 1963). In this paper, it is argued that both reducing and amplifying deviation are the same in that they both focus on the causal loop, and that the only difference is the sign of the feedback in Fig. 7.4. He then corresponds negative feedback to morphological maintenance and positive feedback to morphogenesis.

In the paper, he explains how a farmer comes to a homogeneous plain 1 day and starts cultivating the land, some farmers follow him, eventually one of them opens a farm equipment shop, which becomes a hangout for farmers, and in the meantime a grocery shop is built next door, and so on, using the concept of positive feedback to explain how cities are formed.

Lee Sproull and Sara Kiesler's two-level theory of technological development is precisely the effect of this positive feedback (Sproull & Kiesler, 1992). Sproull and Kiesler argue that the impact of technology appears in two levels: efficiency and social systems. The first stage is the expected increase in efficiency and productivity brought about by the new technology, but then the feedback of the innovation creates unpredictable changes in the system, which lead to the second level of social and organisational change.

There are many examples of this in information technology. According to Sproull and Kiesler, for example, the typewriter, once thought to be used only by ministers and writers, has changed the way office work is done in companies, making typing a safe and respected occupation for young women, producing large volumes of documents and professionalising office work (Sproull & Kiesler, 1992). This concept of feedback has been used in many ways.

This notion of feedback, both positive and negative, has been a key word in subsequent systems theory explanations of self-organisation.

7.6 Structural Similarity

As represented by the feedback structure in cybernetics, it is one of the characteristics of systems theory to focus on the concept of "structure" behind the phenomena. By the way, what is a structure? Mathematically, it is a set with a finite set of operations and a finite set of relations defined on it (Graetzer, 1968).

Let us look at some representative concepts of structure used in various fields. Phonetics considers structure as a set of equivalence relations defined on a set of phonemes. Roman Osipovich Jakobson, who established the structure of phonological systems, said that structure is the interrelationship of features in a given language (Jakobson, 2003). For example, [m] and [ŋ] have the same articulation style, nasal sound and [k] and [g] also have the same articulation style, burst sound.

While [k], [g], [ŋ] have the same articulation point, velum sound.

Those relationships among equivalence relations are called structure in phonetics (Fig. 7.5).

"Systematis Sexualis" of the Linnaean hierarchical classification system proposed by Carl von Linne is based on the equivalence relations of the number of stamen and the shape of the tip of the pistil in a set of plants (Fig. 7.6).

Both of morning glory and gentian have five stamens, while iris has four. Then they are in difference equivalence classes.

From the viewpoint of pistil, the top of pistil of morning glory is not split, while that of gentian is split into two. Then they are included in different equivalence classes.

In this sense, the concept of structure in biology is quite similar as in phonetics.

On the other hand, the well-known concept of structure in the cultural anthropology is "the culinary triangle" by French anthropologist Claude Lévi-Strauss. The triangle is based on the triangle of the "raw", the "cooked" (boiled or baked),

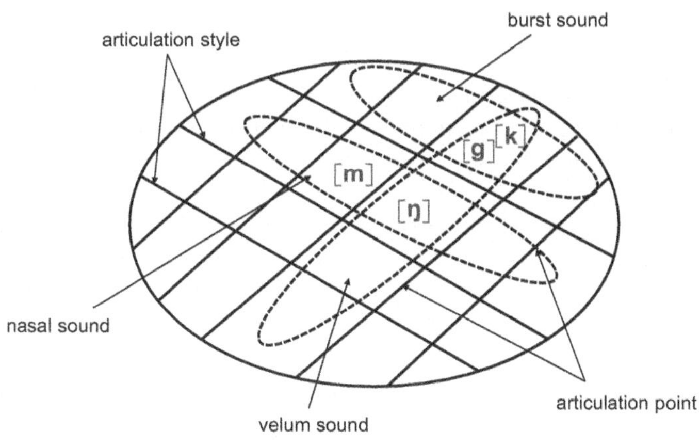

Fig. 7.5 Structure of Phoneme

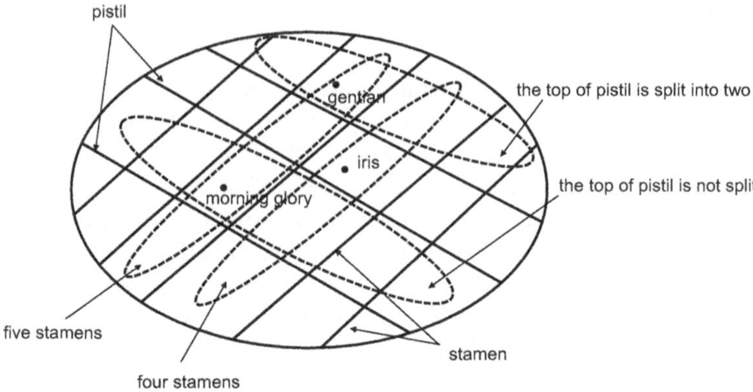

Fig. 7.6 "Systematis Sexualis" of the Linnaean hierarchical classification system

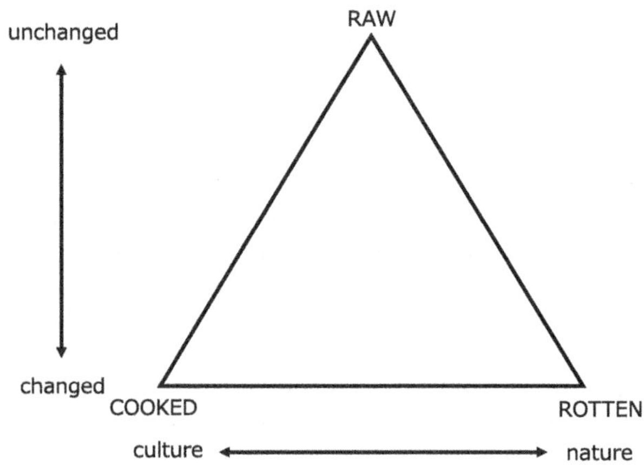

Fig. 7.7 The Culinary Triangle

which is a cultural variant of the "raw", and the "rotten", which is a natural variant of the "raw". The problem here is the relation between the two equivalence relations corresponding to the division "modified"/"unmodified" and the division "culture"/"nature" (Fig. 7.7).

As these examples show, the most common structure is that of a set with a set of equivalence relations defined on it.

There are research to analyse the social customs of human societies using the algebraic structure of groups. One such example is the analysis of the marriage system by Levi-Strauss's.

In the book *Les Structures E'le'mentaires de la Parente'*, Levi Strauss has analysed the marriage system of the Kariera people of Australia (Lévi-Strauss, 1967). There are four classes in this society: A, B, C and D. For each person, there

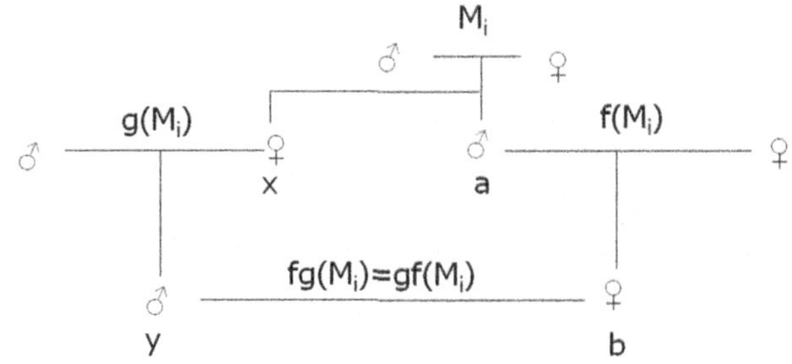

Fig. 7.8 Representation of fg = fg in family tree

is only one type of marriage, which is determined by his/her sex and the type of marriage of his/her parents. That is,

M1: a man of A and a woman of B,
M2: a man of B and a woman of C, and
M3: a man of C and a woman of D,
M4: a man of D and a woman of A.

There are four types of marriage.

If the mother belongs to class A, B, C or D, the child belongs to class B, C, D or A.

Let us illustrate it using Fig. 7.8. Let M = {M1,M2,M3,M4} be the set of four types of marriage. And let f: M → M be the marriage type that a son born in the Mi type of marriage can enter into and g: M → M be the marriage type that a daughter born in the Mi type of marriage can enter into, respectively. It follows from a simple calculation that fg = gf. This implies that the group consisting of {f, g} has the structure called commutative group.

Based on Fig. 7.8, let us consider what this equality means practically. The marriage type of a daughter (x) born in the Mi marriage type is g(Mi), and that of her son (y) is f(g(Mi)). On the other hand, the marriage type of the son (a) born in the marriage type of Mi is f(Mi), and the marriage type of the daughter (b) is g(f(Mi)). Then fg = gf means that both are identical. It means that in the Kariera society, a man can marry the daughter of his mother's brother and a woman can marry the son of her father's sister.

The Mesarovic-Takahara school of mathematical systems theory uses this notion of structure for advanced systems theory. It started with the bold and penetrating view that a system is simply a mathematical relation between a set of inputs X and a set of outputs Y, and gradually added mathematical properties to the system.

Takahara-Ikeshoji finally developed a profound system leading to basic linear systems (Mesarovic & Takahara, 1989).

In the framework of such a mathematical systems theory, the conditions under which a given input-output system can be regarded as consisting of multiple input-output systems are investigated, and this collection of essays also includes a discussion of the conditions under which a subsystem and its constituent subsystems can be regarded as identical.

7.7 Business Dynamics

Now, Systems Dynamics or Business Dynamics is a research field that combines both of the two pillars of systems theory: emergence and structural isomorphism.

In 1956, Professor J. W. Forrester in MIT developed Industrial Dynamics in order to analyse dynamics of firm activity. In 1970, World dynamics has been developed in order to analyse the growth of the world based on the relationship between natural resources and environmental issues after Rome Club conference.

The result has been published in 1972, it is known as *The Limits to Growth* (Meadows et al., 1972). This report caused a worldwide sensation because of its vivid and concrete descriptions of the population explosion and the depletion of food and oil resources that will occur in the near future. This was followed by a second report, *Mankind at the Turning Point*. In this second report to the Club of Rome, the described model represents the Earth as ten mutually exclusive and exhaustive land areas, each represented by a submodel comprising the same six strata in a hierarchical arrangement (Mesarovic & Pestel, 1974). Actually, one of the authors of the book, Professor Mesarović is the supervisor of Professor Takahara when he was studying in the Case Institute of Technology (currently, Case Western Reserve University).

Recently, it is known as Systems Dynamics or Business Dynamics and is being researched centred in MIT (Sterman, 2000).

The most important point of Business Dynamics is that it can clarify causal relationship among variables focusing on feedback loop to grasp a given system as a whole.

Figure 7.9 shows the CLD for the temperature control of an air conditioner and the CLD for the phenomenon of human perspiration. The arrows represent causal links and B means Balance loop. It is easily seen that both CLDs have the same feedback structure with balance loop.

In this sense, the field of Business dynamics handles the two pillars of general systems theory, that is, emergent property and structural similarity.

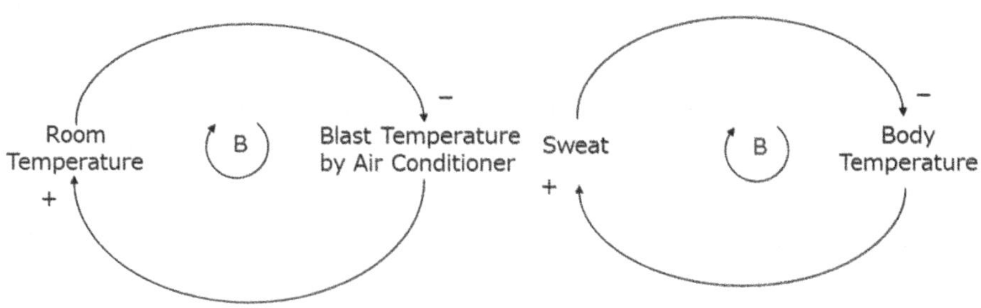

Fig. 7.9 Balanced loop found in Cybernetics

7.8 Soft Systems Methodology

Systems engineering was born and the Apollo mission, which sent mankind to the moon in the 1960s, was said to be a triumph of systems engineering. This approach is known as Hard Systems Methodology because it was aimed at artificial physical systems (hard systems). Hard Systems Methodology has also been applied to solve "wicked problems" but in vain. In the large-scale complex human activity systems such as health care, transportation, urban management, disaster recovery, environmental protection, etc., it is not obvious what the problem is in the first place, and therefore hard systems methodology based on an empirical (logical positivist) approach with a given "objective", such as "sending mankind to the moon", is not suitable. Instead, Checkland developed a phenomenological and hermeneutic approach, which is a search-and-learn type without "purpose" as a given (Checkland, 1981). This is called Soft Systems Methodology.

The Soft Systems methodology starts with a consensus on the purpose of the system, called the root definition. For example, let us consider the design of a system called "prison" (Toda & Iijima, 2000). What is the root definition of "prison"?

The first thing that may come to mind is "a system for educating criminals" or "a system for punishing criminals".

From the point of view of a "system for educating criminals", the next consideration is what kind of education should be provided, but from the point of view of a "system for punishing criminals", the next consideration is what kind of "punishment" is appropriate to impose.

You might see it in terms of "a system that supports the life of the guards" or "a system that allows them to produce things for cheap wages". From these points of view, you will notice that the issues to be considered are quite different. Thus, when we consider a society composed of people with various values, it may be "obvious" that the purpose of the system is not "obvious".

With this background of these debates, a research field called Engineering Systems has been gaining attention since around 2000. Engineering Systems is an interdisciplinary field of engineering, social science and management, which integrates technology, policy, decision analysis, OR (Operations Research), innovation theory, production and management engineering (Fig. 7.10).

7.9 Design Thinking and Design Management

More recently, the approach to problem solving known as "design thinking" has been attracting attention. This is based on the idea that innovation is born at the intersection of "technology", which is concerned with feasibility, "business", which is concerned with sustainability in society, and "people", which is concerned

Fig. 7.10 Engineering Systems

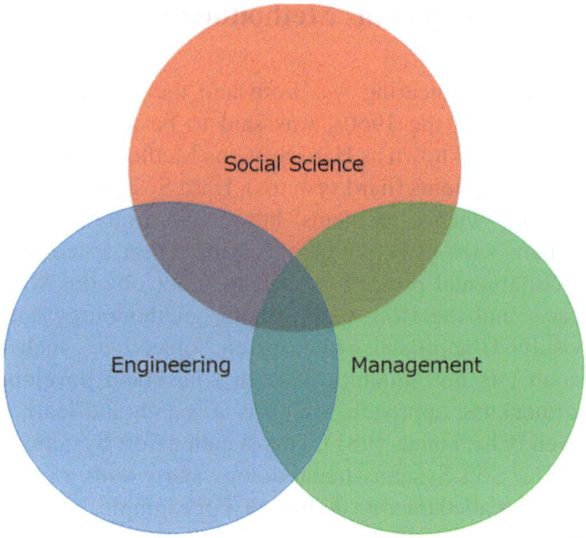

Fig. 7.11 Innovation stems from the intersection

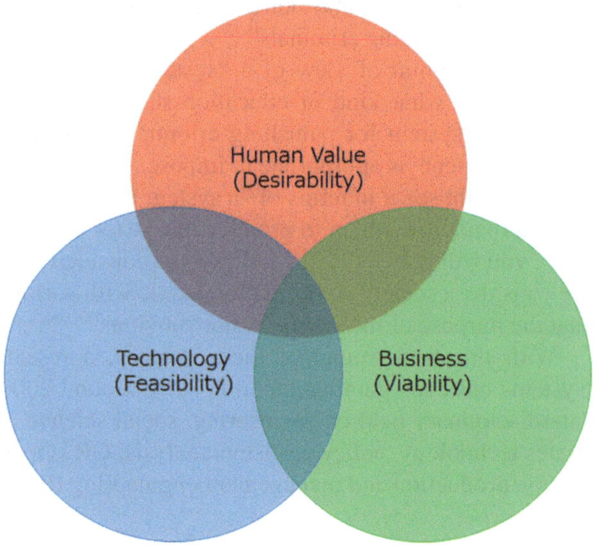

with desirability, as shown in Fig. 7.11. In Engineering Systems, if we think that "engineering" corresponds to "technology", "social science" to "people", and "management" to "business", and then these can be in line with each other.

As mentioned earlier, design is considered here as "the intellectual activity of creating artificial systems", and Jan Dietz considers the general system development process as follows (Dietz & Mulder, 2020):

> The three main phases in a development process are function design, construction design, and implementation design. Function design starts from the ontological model of the using

> system.... Construction design starts from the object system function, and ends with the ontological model of the object system. ... Implementation design starts from the ontology of the object system and ends with the fully detailed specification of a possible implementation....

In this way, Design is the intellectual activity involved in the series of processes of requirements-function-construction-implementation that creates artificial systems supporting to a given problem.

The Danish Design Centre has proposed a concept of maturity in the use of design in management called the Design Ladder (Danish Design Center, 2001) (Fig. 7.12).

Step 1: Non-design: Design is an invisible part of, e.g., product development, and the task is not handled by trained designers. The solution is driven by the involved participants' ideas about good function and aesthetic. The users' perspective plays little or no role in the process.

Step 2: Design as form-giving:Design is viewed exclusively as the final form-giving stage, whether in relation to product development or graphic design. Many designers use the term "styling" about this process. The task may be carried out by professional designers but is typically handled by people with other professional backgrounds.

Step 3: Design as process: Design is not a result but an approach that is integrated at an early stage in the development process. The solution is driven by the

Fig. 7.12 Design Ladder

problem and the users and requires the involvement of a wide variety of skills and capacities, for example process technicians, materials technicians, marketing experts and administrative staff.

Step 4: Design as strategy: The designer works with the company's owners/management to rethink the business concept completely or in part. Here, the key focus is on the design process in relation to the company's business visions and its desired business areas and future role in the value chain.

The ladder here is four steps, but Bucolo extends it by adding the fifth step, Design as organisational transformation, and the sixth step, Design as National Competitive strategy (Bucolo, 2015).

There is also research based on this ladder that seeks to assess the maturity of integrating design into management (Pettigrew et al., 2016).

One of the characteristic ideas of Design Thinking, which is particularly relevant to the third step of this ladder, is to identify problems through empathy with (potential) users. When there was no abundance of products, it worked well to assume that users' needs for products were self-evident and predetermined, but when there is an abundance of products, it is important to understand that different people have different values and that products are created to meet different needs. Design thinking takes this one step further. Design thinking takes this one step further and assumes that users themselves do not know exactly what their needs are. The Scandinavian concept of co-design goes further and does not even use the word "user", but takes the approach of "empathising" with people in an unpleasant situation, thinking together and creating together a story for a new future.

7.10 Conclusions

The title of this essay is "What should we add to science to solve wicked problems?" and I believe that "design" may give us a clue to an answer for this question.

A "wicked problem" is not an object of natural science, but a systemic problem of human activity. Therefore, it is impossible to apply the scientific view represented by the three Rs. Reductionism, Repeatability and Refutation. As for systems theory, based on the failure of hard systems methodology, soft systems methodology has been suggested as a candidate. However, soft systems methodology assumes that the people involved are clearly aware of the objectives and basic definitions of the system, which is not always the case. This is where empathy in "design thinking" becomes important.

Our approach to "wicked problems" is to start from the "wicked problems" of people, organisations and society, identify the problems, create a "blueprint" of an artificial system (an artificial mechanism such as a material product, service or social system) that can solve the problems, and then work with engineers and governments to implement them in society and solve problems. We then work with engineering, administration and other people to implement the solution into society and make the solution sustainable.

Empathy with the target audience is necessary to identify what the problem is and what requirements the solution should meet. On the other hand, in order to develop an artificial system as a concrete deliverable, it is necessary to move away from the black box model, which is a functional representation, to the white box model, which is a constructive representation, i.e., a model concerning "the elements that make up the system, the environmental elements and their interrelationships". Based on the constitutive model, engineering is the process of implementation using concrete techniques. Engineering generates, develops and implements artificial systems as solutions to problems. In order to build a constitutive model from a functional model, we need to know the basic principles of the specific technologies, and to know which technologies should be combined to fulfil a certain function, we need not only knowledge of these technologies but also logical thinking. This is where science comes into play.

Thus, the coexistence of empathy and logic is necessary in the future, and the author believes that in order to solve "wicked problems", we need to add "empathy" to science.

References

Ampère, A.-M. (1834). *Essai sur la philosophie des sciences*. Chez Bachelier.

Buchanan, R. (1992). Wicked problems in design thinking. *Design Issues, 8*(2), 5–21.

Bucolo, S. (2015). *Are we there yet? Insight on how to lead by design*. BIS Publishers.

Camillus, J. C. (2008). Strategy as a wicked problem. *Harvard Business Review, 86*(5), 98–101.

Checkland, P. (1981). *Systems thinking, systems practice*. Wiley.

Checkland, P. (1991). From framework through experience to learning: The essential nature of action research. In H.-E. Nissen, H. K. Klein, & R. Hirshheim (Eds.), *Information systems research: Contemporary approaches and emergent traditions*. Elsevier.

Churchman, C. W. (1967). Wicked problems. *Management Science, 14*(4), B-141–B-142.

Conklin, J. (2009). Building shared understanding of wicked problems. *Rotman Magazine, Winter*, 17–20.

Danish Design Center. (2001). Retrieved from https://issuu.com/dansk_design_center/docs/design-ladder_en

Dietz, J., & Mulder, H. (2020). *Enterprise ontology*. Springer.

Graetzer, G. (1968). *Universal Algebra*. D. van Nostrand.

Jakobson, R. O. (2003). Les Fondations du langage. Essais de linguistique générale I.

Lévi-Strauss, C. (1967). *Les Structures E'le'mentaires de la Parente*. Mouton & Co and Maison des Science de l'Homme.

Maruyama, M. (1963). The second cybernetics: Deviation-amplifying mutual c ausal processes. *American Scientist, 51*(2), 164–179.

Meadows, D. H., Meadows, D. L., Randers, J., & Behrens, W. W., III. (1972). *The limit of growth*. Potomac Associates, Universe Books.

Mesarovic, M., & Pestel, E. (1974). *Mankind at the turning point: The second report to the Club of Rome*. Dutton.

Mesarovic, M. D., & Takahara, Y. (1989). *Abstract systems theory (Lecture Notes in Control and Information Sciences)*. Springer.

Pettigrew, D. K., Thurgood, C., & Bucolo, S. (2016). A design innovation adoption tool for SMEs. In *20th DMI: Academic design management conference inflection point: Design research meets design practice*.

Rittel, H. W. J., & Webber, M. M. (1973). Dilemmas in a general theory of planning. *Policy Sciences, 4*, 155–169.

Roman Osipovich Jakobson. (2003). *Les Fondations du langage*. Essais de linguistique générale I.

Simon, H. A. (1996). *The sciences of the artificial* (3rd ed.). The MIT Press.

Sproull, L. & Kiesler, S. (1992). *Connections: New Ways of Working in the Networked Organization*, The MIT Press.

Sterman, J. D. (2000). *Business dynamics-systems thinking and modeling for a complex world*. Irwin McGraw-Hill.

Suppe, F. (1977). *The Structure of Scientific Theories*. University of Illinois Press.

Toda, Y., & Iijima, J. (2000). *Business process modelling (in Japanese)*. Nikka-Giren Syuppansya.

von Bertalanffy, L. (1968). *General system theory*. George Braziller.

Wiener, N. (1948). *Cybernetics: Or control and communication in the animal and the machine*. Wiley.

Chapter 8
Methodology for Refining Concept Through Refutation in Theory of Organizational Strategy

Ryo Sato

Abstract The general systems research has focused on complexity and provided hierarchical organization model and operation model. The research on corporate strategy can be positioned in higher stratum of the model of organizations. A theory of strategy of organizations has its own concept and propositions, and the proof of a proposition is usually based on case studies. Using case studies as its proof has a reason that experiments are impossible in business phenomena, and that a phenomenon or event is not repeatable.

We propose a methodology of organizational strategy by refining with cases through refutation in this chapter. A universal proposition has both sufficient and necessary conditions. When we are doing research, we can focus on the cases that satisfy necessary condition but may not satisfy the sufficient condition. This attempt of search of cases helps us understand real business activity. In the research area of strategy dynamics, cases are distinct and not many. We need to select cases that are theoretically meaningful and that can be used for refutation-oriented check. It is almost useless that gathering positive cases for an already proved existing proposition. When there is a theory of strategy, business cases should be used for refutation purpose in order to enhance and deepen the scientific knowledge.

Keywords Organization model · Refining concept through refutation · Development of scientific theory of strategy

8.1 Introduction

The general systems theory is a metatheory. One of its aims is to clarify general logical structure of theories of specific systems. With such general structure in

R. Sato (✉)
School of Management, Tokyo University of Science, Tokyo, Japan
e-mail: rsato@rs.tus.ac.jp

K. Kijima et al. (eds.), *Systems Research II*, Translational Systems Sciences 27,
https://doi.org/10.1007/978-981-16-9941-2_8

mind, we can explore complex existence such as human organization and illuminate important issues which are not considered or formulated as problems.

Business is ever active research target. Therefore, general systems research focused on its complexity and provided hierarchical organization model and operation model. The former deals with human decision and its organizational implication, while the latter is used to understand the structure and dynamics of process model and optimization. For instance, I and my colleagues noticed that ambiguity of dataflow diagram to capture business processes in information systems methodologies (Sato & Praehofer, 1997). By applying the concept of state space representation developed in the general systems theory (Mesarovic & Takahara, 1975), we have founded the necessary and sufficient condition for a dataflow diagram to have causality. It is a foundation in the sense that people stated different points in modeling business processes with dataflow diagrams and data models without any firm reason except their experience and that our result shows a reason for the business process to be causal. The result for modeling business process also used in production control mechanism. This was possible because analysis of dynamics characteristics is always possible if we have state space representation of the target system. Kanban system is the name of production control mechanism that has been invented and developed by Ohno Taiichi with Toyota automobile company. A kanban system uses cards to control the flows and accumulation of inventory and production. CONWIP is a variation of kanban systems. CONWIP stands for constant work-in-process. Both types of control systems are quite important in operation of factory to quickly respond to the randomness of car demands with only minimum inventory of parts and row material. Some research showed respective advantages among kanban systems and CONWIP applied for different processed. We have shown how to compare dynamics of those control mechanism and that there is no universal superiority among kanban systems and CONWIP (Sato & Khojasteh, 2012).

The realization theory of dynamical input–output systems has developed by Mesarovic and Takahara (1975, 1989), and the many theoretical applications, including that of dataflow diagrams and kanban system, have used the concepts in the theory. A different type of general model is the goal seeking model, which is adopted when we explore many aspects of complexity in human organizations. For instance, research on corporate strategy can be positioned in higher stratum of the model, and it means strategy decide the lower structure and leads the organizational performance. Since competition among companies across industries are getting fierce, importance of corporate strategy is growing. This is especially true for platform industries (Cusumano, 2010). Mintzberg et al. (2010) showed ten group of disciplines in strategy theory. Since the reality of business and management is quite complex, and since it allows us multifaceted interpretation, theories of strategy inevitably deal with a wide range of issues (Chandler, 1962; Teece et al., 1997; Barney, 2002). A theory of strategy has its own concept and propositions, and the proof of a proposition is usually based on case studies. Using case studies as its proof has a reason that experiments are impossible in business phenomena, and that a phenomenon or event is not repeatable.

Even in this situation, we think that the logic of development of scientific theory is the same among natural and social sciences, according to Popper (1957). And then, by using the same meta-theoretic approach with Mesarovic and Takahara (Mesarovic & Takahara, 1975, 1989; Takahara & Mesarovic, 2003), we propose a methodology of organizational strategy by refining with cases through refutation. In order to develop sound scientific theory of corporate strategy, (1) we need to understand that common strategic or operational measures and policies of high-performance companies are not necessarily sufficient conditions for competitive advantage, and (2) we need to try to find a seemingly contradictory case in the sense that the case seems to satisfy sufficient conditions but does not show high performance.

In this chapter, we will explain as follows. In Sect. 8.2, according to Popper, we discuss the nature of scientific knowledge of natural and social sciences. Refutation of a theory plays a key role to reach scientific proposition. Section 8.3 shows how theory of corporate strategy will be developed by finding business cases for refutation of a theory. In Sect. 8.4, we propose and examine a detailed process of the refining through refutation. We conclude in Sect. 8.5.

8.2 Common Methodology of Natural and Social Sciences

Natural sciences such as physics, chemistry, biology, etc., had brought us with brilliant development of human society with engineering disciplines and technology. Now, we have abundant of artificial and useful things that ease our daily life. Each theory of science has its own concepts and propositions. A proposition is a statement or a description about concepts that generally holds true, where concepts are related to physical objects or phenomena in the real world. Propositions in sciences are also called theorems or laws. In a social science such as economics and theory of business administration, concepts are related to physical or social objects or issues.

Popper (1957) insisted that any science should be developed by the methodology with refutation. A proposition can be refuted by existence of a phenomenon that contradicts the proposition. For instance, in physics, Newton's law of motion of a thing has been accepted as true theory for long time. But later, it gave way to Einstein's special relativity, because there is an observable fact that refutes Newton's law. Like this, scientific theories grow through refutation of existing theory.

According to Popper (1957), social sciences mainly concern explanation of social phenomena, while natural sciences and engineering concern more prediction. In developing explanation, modus ponens and universal propositions are used. When prediction is not important in a social situation like business strategy, we are virtually seeking explanation in scientific way. So, if we have interesting but not-well-explained situation by any existing theory, then we will try to bring or invent new concepts and establish new universal propositions. This kind of scientific research is commonly attempted.

In the development methodology of scientific theory by using refutation, verification and refutation play similar roles for a proposition. Verification of a proposition is to prove its trueness in any cases, so it is impossible if we remember the case that Newton's law was refuted and refines into the special relativity found by Einstein. A law or proposition holds true until it might be shown false in its original statement by the existence of a counter fact that refutes the law. If a proposition is robust against many attempts that try to be refutation, then the plausibility of the proposition's trueness will be maintained. Any science can only be elaborated and developed with attempts of refutation in this way. This doctrine on development of scientific theory is effective for both natural and social sciences in the same way.

Any science focuses on hypothetic propositions that have fruitful implication and are of practical and theoretic importance, and that can be refutable. The way how a proposition comes into mind is not care at all. It is not possible to have any proposition alone without a theory or some concept. We need to have or provide a theoretical framework with concepts and deduced propositions, though it might be intuitive in initial stage of consideration.

Examination of existing theory or new proposal of new concept and theory should try to refute existing ones. Such refutation-oriented new experiments, or application, of cases can examine the validity of existing scientific theory. Not-well-planned examination for refutation is welcome. In fact, an owner of a new restaurant has many propositions to be examined, and she/he will learn through very careful observation about sales, purchase, and inventory without rigorous method. Those issues will be learned by practice, though some theory will be useful for such observation. The activity like this can be regarded as pre-scientific exposition.

8.3 Logical Structure of Development of Theory of Strategy

8.3.1 Universal Proposition and Specific Proposition

The goal of development of natural and social sciences is to discover both concepts and universal propositions so that people can better find and understand meaningful phenomena in real world with the scientific theory. A law can be stated both as a statement and a logical proposition. The latter can be in the form of mathematical theorem, and simply be called a proposition.

There are two basic forms of propositions: a universal proposition and a particular proposition.

A universal proposition is in the following form.

$$(\forall x)\,(P(x) \rightarrow Q(x))$$

Here, x is a logical variable and $P(x)$ and $Q(x)$ stand for the fact that x satisfies properties, P and Q, respectively. The above proposition states that if $P(a)$ is true then $Q(a)$ is true for any a, where a is an arbitrary thing that we concern. By the same token, it is also expressed that a satisfies the condition P, then it satisfies Q. In a universal proposition, $P(x)$ is called the sufficient condition of the proposition, and $Q(x)$ the necessary condition.

A particular proposition is as follows.

$$(\exists x)\,(P(x) \Lambda Q(x))$$

This proposition states that there exists one specific thing a, and it satisfies both properties, P and Q. In other word, this particular proposition expresses that both $P(a)$ and $Q(a)$ are true for existing thing a. Therefore, this statement holds true if and only if a particular thing satisfies the properties, P and Q, at the same time.

When we interpret a universal proposition in management theory, for instance, if $P(x)$ represents that a company, x, outperforms other companies, the $Q(x)$ represents a property or a necessary condition so that the company has accordingly. Since a company is doing business, it aims to perform economically excellent. If a company could not get enough sales or profit, a universal statement on high performance might provide hint for the company to attain good sales or profit, by checking the sufficient condition of the universal statement on the company.

8.3.2 Role of Business Cases in Theory of Strategy as Social Science

Consider a universal proposition in the form of $(\forall x)(P(x) \rightarrow Q(x))$, where $P(x)$ represents that a company, x, shows high performance. In general, any company is interested in conditions to achieve high performance. So, research of strategy with respect to high performance will try to focus on characterization of sufficient conditions for high performance, which are useful and effective practices for business. Furthermore, such research needs to make clear the relation to existing theories of strategy such as resource-based view or dynamic capabilities.

Let us consider kanban system, for example. Kanban system is also known as lean production system, or the Toyota system, though kanban system is part of lean or Toyota system. The idea and implementation of pull system with kanban was started by Mr. Ohno of Toyota in 1948 (Ohno, 1988). Since the concept of pull system was so revolutionary that developing it over an automobile factory and its suppliers was not easy or simple. It took more than 10 years to implement in the factory. The lean production system is quite effective to operate multi-products-and-variable-production with less or minimum inventory of part, material, and finished products. The lean system is still developing even now, corresponding to current technology and global environmental change.

Now, assume that you, a strategy researcher, could have a chance to talk with Mr. Ohno. Then, you might get deep impression from him and the factory with neat workers. You would focus on the logic of production and inventory control, and also would find some organizational issues that include management of suppliers. When your research had reached to summarizing findings, the conclusion can be said as follows.

"Successful automobile manufacturing companies adopt a pull system with kanban mechanism and very short setup times of respective machines. They attained high turnover of inventory and high quality, and then enjoyed big profit margin." Or, more simply, it is expressed that "if an automobile manufacturer is of high performance, then it uses a kanban system."

Let $K(x)$ mean that a company x operates with a kanban system. Suppose that you got the result for any company, x, in the automobile industry that if x adopts a kanban system then x shows high performance, which is represented as $P(x)$. The statement can be expressed as $(\forall x)(P(x) \leftrightarrow K(x))$.

This proposition has practically useful. If a person of a company happened to see this proposition, then she/he will examine the effect of introducing kanban system into the factory. If a similar concept was already applied for a factory, but did not show sufficient excellence yet, then further measures will be sought through careful examination.

Roos et al. (1990) named the Toyota system as lean system. They used the word, lean system, because their research showed that effectiveness of the control mechanism and manufacturing organization of Toyota system was implemented in some factories in the USA, Europe, and South America. It is not dependent to Japanese culture.

Though a survey of certain number of factories and companies has surprising and practically important implication, there might be a chance that the survey missed a company with kanban system and lower performance. If the surveyed companies are those with kanban systems and high performance, then the resultant proposition will be $(\forall x)(K(x) \wedge P(x) \leftrightarrow K(x))$. Apparently, it is tautology, and so, does not add any knowledge.

8.4 Finding Possible Characteristics for Necessary and Sufficient Conditions

Suppose that we are seeking validity of the proposition such that if a company shows high performance, then it uses a kanban system.

1. In this proposition, a kanban system is in the necessary condition. Since we aim to find a sufficient condition to have high performance, we wonder that adoption of a kanban system will imply high performance. A question that we must clarify is that a kanban system brings us with many good operational practices of manufacturing and resultantly high performance. That is, we will try to find a company, c, in Fig. 8.1 that does not achieve high performance with a kanban system.

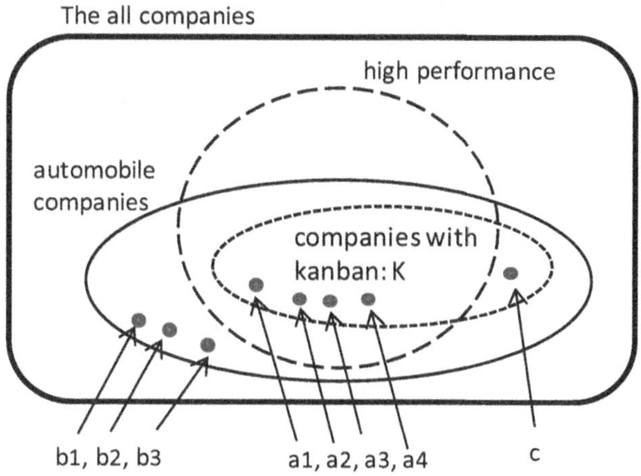

Fig. 8.1 A situation of relation between kanban systems and performances

If such company c is operating its business, we investigate that company's operation and try to make clear the reason of its ordinary or weak performance. We will analyze c's strategy, business process with its kanban system, cost structure and other issues that prohibit high performance.

2. In search of necessary and sufficient conditions for high performance, we try many possible factors and issue. Especially, we want to see whatever important and what the others did not recognized as key factors for high performance. When we seek original and unknown-so-far factors, we need to recognize the pitfall to be avoided.

So, let us consider the following fictitious situation. Assume that we are doing research in automobile industry, and that we could have special reservation to interview the respective presidents and key persons of four companies, a_1 through a_4, in Fig. 8.1. One of our astonishing findings about those four was that each of them has been rearing tigers in headquarters respectively. We analyzed keenly and reached a tentative conclusion that rearing tigers gives board members strong stimuli in motivation. This seemed to be a hidden and deep information for company's high performance. Now, let us denote as $T(x)$ to represent a statement that x rears tigers. Now, we have the following proposition:

$$(\forall x)\,(P(x) \rightarrow [K(x) \Lambda T(x)])$$

Assume also that we took further research on those four companies, and that those four adopted the defined contribution pension system, which leads the four the competitive knowledge of flexible assignment human resources by recruiting wide variety of personnel. Now, by denoting as $D(x)$ to represent that x adopted the defined contribution pension system, our hypothetic proposition can be refined as

$$(\forall x)\,(P(x) \rightarrow [K(x) \Lambda T(x) \Lambda D(x)])$$

Like this, by employing this refinement procedure, any common issues on the four companies, a_1 through a_4, can be added in the necessary condition with the conjunctive connector, "$\wedge$". We understand that rearing tigers in headquarters has no relation to high performance. But, as like $D(x)$ and $T(x)$, whatever factors, that are common for the four companies, can be included in the proposition. The reason that rearing tigers should not be included, usually comes from different knowledge. For instance, knowledge of operations management or usual business practice can be used to eliminate such wrong factors through observation and examination of the real world.

Furthermore, if we examine the propositions in the form, $(\forall x)(P(x) \rightarrow Q(x))$, where $P(x)$ means that x shows high performance, and if we are focusing on the four companies, a_1 through a_4, in Fig. 8.1, then any fact about companies, b_1 through b_4, that do not show high performance, has no influence to the trueness of the propositions with $P(x)$ as the sufficient condition. The company, b_1, may rear a tiger, and b_2 not. The company, b_3, may satisfy $D(b_3)$ and $\neg T(b_3)$, meaning that b_3 employs the defined contribution pension system and does not rear tigers. Such situation on b's does not affect to the trueness of the proposition, $(\forall x)(P(x) \rightarrow Q(x))$.

Breaking seemingly commonsense knowledge is exciting in science. Refining Newton's law (a proposition) of motion by Einstein's special relativity is a striking example. In social science, especially in business strategy, the situation is not so crystal clear, but still be surprising. Ogawa and Mizuno (2004) is an example in the analysis of the success of convenience stores in Japan. Japan has several convenience-store franchises. The top is seven-eleven, and it excels others in the sense that a franchisee shop of the seven-eleven sells around 30% more than other brands of convenience franchisers in Japan for more than 40 years. Anyone wants to see why. Many researchers pointed seemingly plausible characteristics in operation, organization, supply chain structure, information systems for ordering, etc. By thorough research, only Ogawa and Mizuno could have reached the fact that only the seven-eleven has its devoted supplier for fast food such as Japanese lunchbox. What they have done is that refinement of a proposition that states characteristics of high performance in competing convenience-store business in Japan.

As a researcher, we would like to articulate a sufficient condition, $R(x)$, for high performance, $P(x)$, so that a company can implement R for P. That is, a target proposition that we seek is

$$(\forall x)\,(R(x) \rightarrow P(x))\,.$$

As a usual research activity, we try to find characteristics of companies with high performance. It can lead propositions of the form, $(\forall x)(P(x) \rightarrow Q(x))$. By doing so, we will hopefully get closer to a necessary and sufficient condition, $S(x)$, such that we have $(\forall x)(P(x) \leftrightarrow S(x))$. In mathematics exercise, this way of pursuit of sufficient condition from necessary conditions is often used. Because a necessary condition sometimes gives us with good clue for sufficient conditions, researchers of strategy also find the necessary conditions applicable and useful.

In the process of searching a sufficient condition for high performance, $P(x)$, it is also useful for us to try to show that the seemingly sufficient condition is not the case. It is useful because such consideration possibly gives us interesting or important information. For a proposition, $(\forall x)(R(x) \rightarrow P(x))$, the fact that $R(x)$ is not a sufficient condition is proved by showing that the negation of the proposition is false. That is, the negation is false if and only if $(\exists x)(R(x) \Lambda \neg P(x))$ holds true. Or, if there exists a company, e, such that $R(e)$ and $\neg P(e)$ hold true, then the negation holds true.

Let us back to Fig. 8.1 and consider a proposition, $(\forall x)(K(x) \rightarrow P(x))$. It states that if a company employs a kanban system, then it is high performing. For the company, c, in Fig. 8.1, $K(c)$ and $\neg P(c)$ hold true. When the company, c, exists, then a kanban system is too wide condition for an automobile company to be high performance. So, further conditions (e.g., good operations practice) may be required to show high performance. Exploring such additional conditions propel the development of theory of strategy.

We can put the situation simply as follows. Assume that we have a hypothesis that states $(\forall x)(P(x) \rightarrow Q(x))$, and that we reached the hypothesis through considerable research effort such as interviews, data search in many areas, data processing and modeling like SEM for many months or years. How can we generalize this result? Some authors of papers point out at its discussion or conclusion of a paper that in future research the authors would apply the resultant statistical proposition to different industries, expecting that the proposition holds true. But, based on the consideration of this chapter, trying to find a company, z, such that $\neg P(z)$ and $Q(z)$ hold true, will be fruitful and strengthen the paper's conclusion. Because if it is difficult to find any counter example for the proposition, then the plausibility of the proposition increases, and also, we may come closer to the necessary and sufficient condition. Popper's development-by-refutation methodology has such implication for both practice business strategy and theory development.

8.5 Conclusion

In this chapter, we proposed and examined a methodology of for refining theory through refutation. The result of this chapter is summarized as follows.

1. Popper insisted that the natural and social sciences have the same dynamic development methodology as a science (Popper, 1957). By using the universal propositions and individual observations, the methodology brings us with new knowledge through inference mechanism of modus ponens. Furthermore, any scientific theory should allow us refutation-oriented observations with cases and new concepts. Such cases might refute existing paradigm with theory and concepts, and then, facilitate development of scientific theories, refutation in strategy research is sometimes not in simple format like just a negation. Miller and Shamsie (1996), for instance, resolved a complicated historical development

of American film industry by introducing new concepts, and analyzed it with the concepts. They have virtually enhanced the resource-based view of corporate strategy, and then, refuted the existing theoretic framework.

Tanaka and Sato (2016) analyzed the development of dynamic capability of the companies in online game industry in Japan. They focused on the relation among SNS platforms and vendors of online games during 2006 and 2012. They proposed that each of the highly performing SNS platforms had the ability of sensing promising game vendors as the complementors of a platform, and deeply integrated with the vendors. They used three companies' cases and refined through refutation of existing theories of platform strategy.

2. Business cases can be used like Miller and Shamsie (1996) and Tanaka and Sato (2016). Moreover, we can use cases when we examine the plausibility of hypothetic universal propositions. A universal proposition has both sufficient and necessary conditions. When we are doing research, we can focus on the cases that satisfy necessary condition but may not satisfy the sufficient condition. This attempt of search of cases helps us understand real business activity. In the research area of strategy dynamics, cases are distinct and not many. So, we need to select cases that are theoretically meaningful and that can be used for refutation-oriented check. It is almost useless that gathering positive cases for an already proved existing proposition. This issue on cases has been pointed out by Popper. When there is a theory of strategy, business cases should be used for refutation purpose in order to enhance and deepen the scientific knowledge.

It is our strong belief that the methodology of this chapter is useful in the exposition of Mesarovic and Takahara's general cybernetic model of organization.

References

Barney, J. B. (2002). *Gaining and sustaining competitive advantage* (2nd ed.). Pearson Education.

Chandler, A. D., Jr. (1962). *Strategy and structure*. Massachusetts Institute of Technology.

Cusumano, M. A. (2010). *Staying power*. Oxford University Press.

Mesarovic, M., & Takahara, Y. (1975). *General systems theory: Mathematical foundations*. Academic Press.

Mesarovic, M., & Takahara, Y. (1989). *Abstract systems theory*. Springer.

Miller, D., & Shamsie, J. (1996). The resource-based view of the firm in two environments: The Hollywood film studios from 1936 to 1965. *The Academy of Management Journal, 39*(3), 519–543.

Mintzberg, H., Ahlstrand, B., Lampel, J. B. (2008). Strategy Safari, 2nd Edt. Pearson Education Canada.

Ogawa, S., & Mizuno, M. (2004). "Verifying the myth of convenience-stores" organizational. *Science, 37*(4), 52–63. (written in Japanese).

Ohno, T. (1988). *Toyota production system- beyond large-scale production*. CRC. (The original Japanese edition was published in 1978).

Popper, K. (1957). *The poverty of historicism*. Routledge and Kegan Paul.

Roos, D., Womack, J., & Jones, D. (1990). *The machine that changed the world*. Scribner.

Sato, R., & Khojasteh, Y. (2012). An integrated framework for card-based production control systems. *Journal of Intelligent Manufacturing, 23*(3), 717–731.

Sato, R., & Praehofer, H. (1997). A discrete event model of business system—A systems theoretic foundation for information systems analysis: Part 1. *IEEE Transactions on Systems, Man, and Cybernetics, 27*(1), 1–10.

Takahara, Y., & Mesarovic, M. D. (2003). *Organization structure: Cybernetic systems foundation*. Springer.

Tanaka, A., & Sato, R. (2016). Dynamic capability of social game industry in Japan: An analysis of platform formation from 2006 to 2012. *Journal of Japanese Operations Management and Strategy, 6*(1), 55–69.

Teece, D., Pisano, G., & Shuen, A. (1997). Dynamic capabilities and strategic management. *Strategic Management Journal, 18*(7), 509–533.

Chapter 9
PVaR: A New Risk Measure for Financial Investments

Chunhui Xu

Abstract Most risk indicators for financial investments are for the cases with a fixed investment term; they become inappropriate when the investment term is not fixed, which is common in real investment situations. In order to measure market risk of investments in such common situations, we proposed the notion of Period Value at Risk(PVaR) in early 2010s. This article is to introduce PVaR and its computation methods that we developed in recent years.

Keywords Financial investment · Risk measure · Value at risk (VaR) · Period VaR · Stochastic process · Historical simulation · Monte Carlo simulation

9.1 Risk Measures for Financial Investments

Risk is one of the key factors to consider in financial investments, and the measure of risk is the base for risk management. The first measure for financial risk was proposed in Markowitz (1952), which used variance as a measure for risk. This risk measure has been very popular since the 1950s and is the cornerstone of modern investment theories. Despite its popularity, variance is not a satisfactory measure of risk for two reasons: it penalizes gains and losses in the same way, and it is not directly connected to loss. To measure financial risk more properly, many other measures of risk have been proposed. These include semivariance and absolute variance (Konno and Yamazaki 1991); lower partial moments (Bawa and Lindenberg 1977); Value at Risk (VaR, RiskMetrics Group 1996); and maximum loss (Young 1998). While most of these risk measures were developed to measure downside risk, some were developed to improve computational efficiency in solving investment models with risk included. Refer to Dowd (2002) for a survey of market risk measures.

C. Xu (✉)
Department of Risk Science in Finance and Management, Chiba Institute of Technology, Chiba, Japan
e-mail: xchunhui@yahoo.co.jp

K. Kijima et al. (eds.), *Systems Research II*, Translational Systems Sciences 27,
https://doi.org/10.1007/978-981-16-9941-2_9

The VaR proposed by J.P. Morgan was supported by many financial institutes including the Bank for International Settlements. It has become a standard risk measure in the financial industry for its conceptual simplicity and practical application. Refer to the study by Choudhry and Tanna (2006) for a detailed introduction to VaR. However, VaR is not subadditive (Artzner et al. 1999) and does not indicate the size of the potential loss when the loss exceeds VaR. To remedy these shortcomings, several variations of VaR were developed, such as worst conditional expectation (Artzner et al. 1999; Benati 2004) and conditional Value at Risk (CVaR, Rockafellar and Uryasev 2000), which is also called expected shortfall (ES) or expected tail loss (ETL) (Acerbi and Tasche 2002; Tasche 2002).

However, most measures for risk including VaR and its variations reflect the risk at a certain future time, which can be seen from the methods to compute these risk measures. In computing market risk measured by any conventional risk indicators, risk factors are regarded as stochastic variables, such as taking the price of a certain stock at some future time as a stochastic variable; market risk is measured by the stochastic features of the risk factors. The values of risk factors till the future time have not been taken into account in the computation, hence risk estimated in this way is only the risk at some future time, it does not reflect the risk during a period of time, and it cannot be used directly when the investment term is not fixed.

In real investment situations, most investors may not have a clear idea regarding the time they will sell the financial instruments in hand; they may anticipate a time span they may hold the financial instruments. However, this does not mean that they have to sell the instruments at the end of the time span; they are usually flexible regarding the termination time of the investment, hence they care about the risk over the whole investment time span, not just about the risk at the end of the time span. In addition, some fund managers dispose any financial instruments when their depreciation exceeds some predetermined threshold say 50%. Hence, there is uncertainty regarding the liquidating time in many investment situations, the risk within a time span is important for investors. However, risk indicators up to date fail to reflect the risk involved over a period of time.

To measure the risk of investments over a period of time, I extended the concept of VaR to Period Value at Risk(PVaR), as a new risk indicator in early 2010s. PVaR measures the market risk when the investment term is not fixed but within a period of time, a common investment situation in reality.

This article is to introduce PVaR and the methods to compute it. The remainder of this article is organized as follows. Section 9.2 proposes and formulates the notion of PVaR. Section 9.3 considers the calculation of PVaR by focusing on a special case: there is only one risk factor in investments. We develop an analytic expression for PVaR on the basis of the assumption that the risk factor can be expressed as a geometric Brownian motion. Section 9.4 introduces a method to estimate PVaR when the values of risk factors can be simulated by the historical simulation method. Section 9.5 introduces a method to estimate PVaR when the values of risk factors can be simulated by the Monte Carlo simulation method. Section 9.6 gives an example to illustrate one numerical method. Finally, conclusions and comments are presented in Sect. 9.7.

9.2 Notion and Formulation of Period Value at Risk

Without losing generality, we explain and formulate the notion of PVaR by focusing on a simple situation in which there is only one risk factor in investments.

Denote the length of longest holding period of the investment by T_{max}, with t_a as the beginning and t_b as the end of the possible liquidation period. That is, the investment may be liquidated at a time between t_a and t_b. For the sake of brief notation, we use 0 and T to indicate the beginning and end of the possible liquidation period of the investment, respectively.

Let $\omega(t)$ be the value of the risk factor at time t, which may be regarded as the price of the risk factor at time t, and $\omega(t) : t \in [0, T]$ is supposed to be a stochastic process within $[0, T]$.

As a risk indicator of an investment, it should be related directly or indirectly to the loss of the investment. Denote the investment by x, and the loss rate of x when liquidated at time t by $L(x, \omega(t))$, then the maximum loss rate of investment x when the liquidation is within time span $[0, T]$ is calculated as follows,

$$L_T(x, \omega) = max\{L(x, \omega(t)) : t \in [0, T])\}. \tag{9.1}$$

Obviously, $L_T(x, \omega)$ is a stochastic variable. The greatest loss rate that may occur at probability $1 - \alpha$, denoted by d^{min}, is given by

$$d^{min} = inf\{d \in R^1 | P(L_T(x, \omega) \geq d) \leq 1 - \alpha\}. \tag{9.2}$$

We define d^{min} as the Period Value at Risk with confidence level α, denoted by $PVaR_\alpha(x)$.

Definition 1 The Period Value at Risk (*PVaR*) at confidence level α of an investment x is defined in the following formula:

$$PVaR_\alpha(x) = inf\{d \in R^1 | P(L_T(x, \omega) \geq d) \leq 1 - \alpha\}. \tag{9.3}$$

$PVaR_\alpha(x)$ is the greatest loss rate that may occur with probability $1 - \alpha$ when the liquidation is within a period of future time. The meaning of PVaR is illustrated in Fig. 9.1.

Remark 1 Taking $\omega(t)$ in formula (9.1) as a vector of risk factors, we can see that the above definition of PVaR applies to cases in which there are many risk factors; to save space, we do not rewrite the formulation.

To see the difference with conventional VaR, we state the definition of VaR as follows. The Value at Risk with confidence level α of investment x is defined in the following formula:

$$VaR_\alpha(x) = inf\{d \in R^1 | P(L(x, \omega(T)) \geq d) \leq 1 - \alpha\}, \tag{9.4}$$

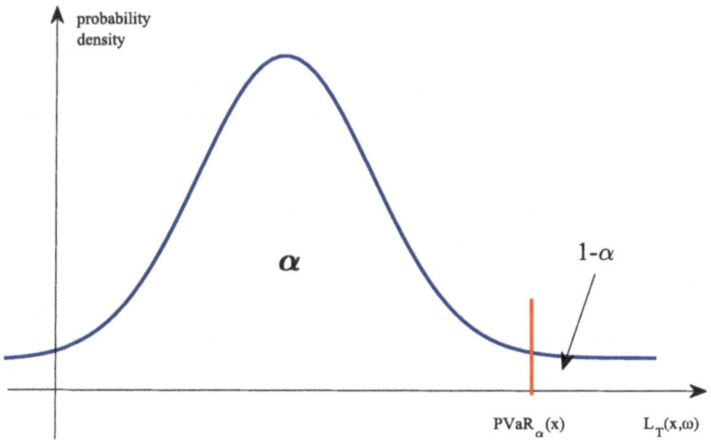

Fig. 9.1 PVaR: Measure of market risk of an investment in a period of time

where $L(x, \omega(T))$ is the loss rate at time T. Hence, VaR reflects the risk at a certain future time, while PVaR is related to the risk over a future time span.

9.3 Calculation of PVaR

Because PVaR is an indicator of investment risk over a period of time in the future, some sort of information about the risk factor during this time period is needed to calculate PVaR.

This section assumes that the values of risk factor form a specific stochastic process and calculates the PVaR of an investment based on this assumption.

Assumption 1 *The values of the risk factor can be modeled as a geometric Brownian motion. That is, the value of the risk factor, denoted by $\omega(t)$, is the solution of the following equation,*

$$d\omega(t) = \mu\omega(t)dt + \sigma\omega(t)dB(t), \tag{9.5}$$

where μ, σ are parameters describing a geometric Brownian motion, and $B(t)$ is the standard Brownian motion.

Refer to Karatzas and Shreve (1991) for an introduction to Brownian motion. We derive an analytic formula to calculate PVaR under this assumption.

Define the profit rate of such an investment, denoted by $R(\omega(t))$, as the logarithmic return, i.e.,

$$R(\omega(t)) = ln\frac{\omega(t)}{\omega(0)}, \tag{9.6}$$

where $\omega(0)$ is the purchase price of the risk factor.

Then we can easily show that the profit rate of such an investment under Assumption 1 is given by

$$R(\omega(t)) = (\mu - \frac{\sigma^2}{2})t + \sigma B(t). \tag{9.7}$$

Define loss rate as the negative profit rate. Then, the loss rate of an investment under Assumption 1 is given by

$$L(\omega(t)) = -(\mu - \frac{\sigma^2}{2})t - \sigma B(t). \tag{9.8}$$

Because $L(\omega(t))$ given by (9.8) is a Brownian motion with drift, we will make use of the properties of Brownian motion to calculate PVaR. We can prove two conclusions related to Brownian motion with drift, as stated in the following two lemmas.

Lemma 1 *Let $X(t)$ be a linear Brownian motion with drift: $X(t) = at + B(t)$, where B(t) is the standard Brownian motion and a is a constant parameter. Denote the maximum value of $X(t)$ within $[0, T]$ by X_T: $X_T = max_{0 \leq t \leq T} X(t)$. Then the joint distribution function of X_T and $X(T)$ is given as follows:*

$$G(x, y) = P(X_T \leq x, X(T) \leq y) = \Phi(\frac{y - aT}{\sqrt{T}}) - exp(2ax)\Phi(\frac{y - 2x - aT}{\sqrt{T}}), \tag{9.9}$$

where $\Phi(\cdot)$ is the cumulative distribution function of the normal distribution.

Define another linear Brownian motion with drift as follows:

$$Y(t) = at + \sigma B(t). \tag{9.10}$$

Let $Y_T = max_{0 \leq t \leq T} Y(t)$; then the joint distribution function of Y_T and $Y(T)$ can be calculated as follows:

$$F(x, y) = P(Y_T \leq x, Y(T) \leq y) = P(\frac{Y_T}{\sigma} \leq \frac{x}{\sigma}, \frac{Y(T)}{\sigma} \leq \frac{y}{\sigma}).$$

Because $\frac{Y(t)}{\sigma} = \frac{a}{\sigma}t + B(t)$ holds, Lemma 1 implies that the joint distribution function of Y_T and $Y(T)$ is given as follows:

$$F(x, y) = \Phi(\frac{\frac{y}{\sigma} - \frac{a}{\sigma}T}{\sqrt{T}}) - exp(\frac{2a}{\sigma}\frac{x}{\sigma})\Phi(\frac{\frac{y}{\sigma} - 2\frac{x}{\sigma} - \frac{a}{\sigma}T}{\sqrt{T}})$$
$$= \Phi(\frac{y - aT}{\sigma\sqrt{T}}) - exp(\frac{2ax}{\sigma^2})\Phi(\frac{y - 2x - aT}{\sigma\sqrt{T}}).$$

Hence, we have the following conclusion.

Lemma 2 *Let $Y(t) = at + \sigma B(t)$ be a linear Brownian motion with drift, and $Y_T = max_{0 \le t \le T} Y(t)$; then the joint distribution function of Y_T and $Y(T)$ is given as follows:*

$$F(x, y) = P(Y_T \le x, Y(T) \le y) = \Phi(\frac{y - aT}{\sigma\sqrt{T}}) - exp(\frac{2ax}{\sigma^2})\Phi(\frac{y - 2x - aT}{\sigma\sqrt{T}}). \tag{9.11}$$

The loss rate given in (9.8) is a linear Brownian motion with drift. We can get the cumulative distribution function of L_T, the maximum loss rate in period $[0, T]$, as stated in the following proposition.

Proposition 1 *When the value of a risk factor can be modeled as a geometric Brownian Motion, as stated in Assumption 1, the cumulative distribution function of the maximum loss rate in period* $[0, T]$, *denoted by* $F(x)$, *is given by*

$$F(x) = P(L_T \le x)$$
$$= \Phi(\frac{x + (\mu - \sigma^2/2)T}{\sigma\sqrt{T}}) - exp(\frac{-2x(\mu - \sigma^2/2)}{\sigma^2})\Phi(\frac{-x + (\mu - \sigma^2/2)T}{\sigma\sqrt{T}}). \tag{9.12}$$

Let the inverse function of F be F^{-1}. We know from Definition 1 that PVaR at confidence level α is $F^{-1}(\alpha)$, i.e.,

$$PVaR_\alpha = F^{-1}(\alpha). \tag{9.13}$$

Therefore, $PVaR_\alpha$ is the solution of the following equation with x as the variable:

$$\Phi(\frac{x + (\mu - \sigma^2/2)T}{\sigma\sqrt{T}}) - exp(\frac{-2x(\mu - \sigma^2/2)}{\sigma^2})\Phi(\frac{-x + (\mu - \sigma^2/2)T}{\sigma\sqrt{T}}) = \alpha. \tag{9.14}$$

Solving Eq. (9.14) analytically is hard; however, we can numerically determine the solution by using some commercial software like Matlab, a numerical computing environment developed by MathWorks.

Table 9.1 Computation results from solving Eq. (9.14) ($\mu = 0.3$ and $\sigma = 0.2$)

α	0.80	0.82	0.85	0.87	0.90	0.92	0.95	0.97	0.99
$PVaR_\alpha$	0.1092	0.1161	0.1280	0.1373	0.1541	0.1683	0.1976	0.2286	0.2926

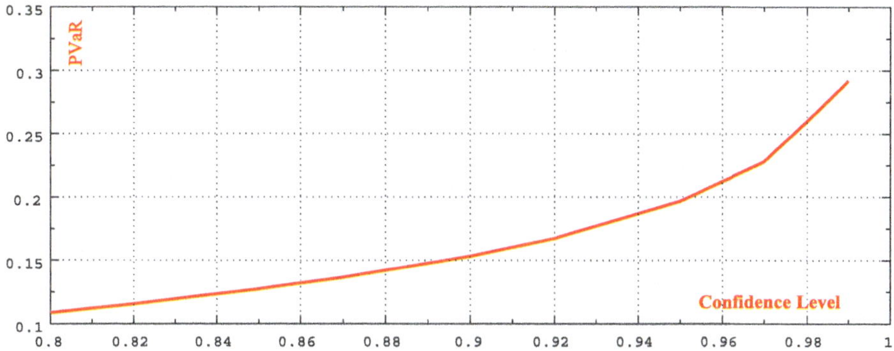

Fig. 9.2 Relation between PVaR and its confidence level

We employ MATLAB in solving Eq. (9.14) for different confidence levels by setting $\mu = 0.3$ and $\sigma = 0.2$. The results are summarized in Table 9.1. Table 9.1 is also plotted in Fig. 9.2. As anticipated, PVaR monotonically increases as the confidence level of PVaR increases.

When there are several risk factors in an investment or the risk factors cannot be described as the geometric Brownian motion, computing the PVaR of an investment with the analytical method turns out to be difficult. We will introduce numerical methods to estimate PVaR in the following sections.

9.4 Estimation of PVaR Using the Historical Simulation Method

This section introduces a numerical method for estimating PVaR, which was proposed in Xu et al. (2012). This method assumes that the values of risk factors in the future can be generated via the historical simulation method.

Assumption 2 *The values of risk factors in the future can be simulated with the historical simulation method.*

Suppose we have produced scenarios using the historical simulation method for the stochastic process $\omega(t) : t \in [0, T]$, each scenario consisting of the values of the stochastic process at m different times: $t_1, \cdots, t_m$, denote the values of the stochastic process in the ith scenario by

$$\omega^i(t_j), j = 1, 2, \cdots, m, t_j \in [0, T].$$

The loss rate of an investment can usually be calculated when the values of risk factors in the future are known. Denote the loss rate in the ith scenario at time t_j by $L(x, \omega^i(t_j))$, then the maximum loss rate in the ith scenario, denoted by $L_T^i(x)$, is given by

$$L_T^i(x) = max\{L(x, \omega^i(t_j)) : j = 1, 2, \cdots, m\}. \tag{9.15}$$

Let N be the number of simulated stochastic processes, and denote the kth smallest value in the set $\{L_T^i(x), i = 1, 2, \cdots, N\}$ by $L_T^{\{k\}}(x)$.

If k is set to be the largest integer among the integers smaller than $N\alpha + 1$, such a value is to be denoted by $\lceil N\alpha \rceil + 1$; then the proportion of the loss rates greater than $L_T^{\{k\}}(x)$ in $\{L_T^i, i = 1, 2, \cdots, N\}$ will not be greater than $1 - \alpha$. According to the definition of PVaR, $L_T^{\{k\}}(x)$ is PVaR at confidence level α, i.e.,

$$PVaR_\alpha(x) = L_T^{\{\lceil N\alpha \rceil+1\}}(x). \tag{9.16}$$

To clearly illustrate the above method, we fill the data of the loss rate into a table, with each row filled with loss rates in each scenario and the rightmost column filled with the greatest data in each row, as shown in Table 9.2. Then PVaR at confidence level α is the $(\lceil N\alpha \rceil + 1)$th smallest value in the rightmost column.

Hence, if scenarios for the risk factor in the possible liquidation period can be produced, we can numerically estimate PVaR of an investment, as stated above.

Remark 2 The above simulation-based method can be extended in a straightforward manner to cases in which many risk factors are involved by letting $\omega(t)$ be a vector of risk factors; hence, we can numerically estimate PVaR by using formulas (9.15) and (9.16) when the scenarios of the risk factors in the possible liquidation period are produced.

Remark 3 The above method implicitly assumes that the liquidating time is uniformly distributed over the possible liquidating period. When the liquidating probability is different at some times, we can adapt the above method to this situation by replacing the loss rate with the expected loss rate.

Let $p(t_j)$ be the liquidating probability at t_j, and $\bar{L}(x, \omega^i(t_j))$ be the expected loss rate, then we have the following relation.

$$\bar{L}(x, \omega^i(t_j)) = p(t_j)L(x, \omega^i(t_j)). \tag{9.17}$$

Table 9.2 Estimation of PVaR with Historical Simulation

	$L(x, \omega^i(t_1))$	$L(x, \omega^i(t_2))$	$\cdots$	$L(x, \omega^i(t_m))$	$L_T^i(x) = max\{L(x, \omega^i(t_j)) : j = 1 \sim m\}$
ω^1			$\cdots$		
ω^2			$\cdots$		
$\vdots$	$\vdots$	$\vdots$	$\cdots$	$\vdots$	$\vdots$
ω^N			$\cdots$		

9.5 Estimation of PVaR Using the Monte Carlo Simulation Method

As shown in the previous section, estimating PVaR with the historical simulation method is easy when the scenarios for the risk factors are obtained. The problem with this method is that the estimation error is unknown, it can not tell the quality of its estimation.

Taking the estimation made by this method as one sample of PVaR, this method obtains one sample of PVaR and uses this sample as an estimation of PVaR, it is the reason why its estimation error is unknown.

If we can produce a series of samples of PVaR and use these samples to estimate PVaR, we can describe the quality of the estimation, this idea leads to a different method for estimating PVaR with Monte Carlo simulation, which was proposed in Huo et al. (2021), this section will present this method.

Suppose we can generate scenarios for the risk factors using Monte Carlo simulation and use the method described in the previous section to get one sample of PVaR with the scenarios generated, repeating this process will produce a series of samples for PVaR.

To estimate PVaR this way, we assume that the values of risk factors in the future can be generated via Monte Carlo simulation, that is,

Assumption 3 *The values of risk factors of an investment in the future can be simulated with Monte Carlo simulation.*

Denote the lth data set of N scenarios for the stochastic process $\omega(t) : t \in [0, T]$ generated with Monte Carlo simulation by A^l, that is,

$$A^l = \{\omega^i(l, t_j), j = 1, 2, \cdots, m; i = 1, 2, \cdots, N\}, \tag{9.18}$$

where $\omega^i(l, t_j)(j = 1, 2, \cdots, m)$ are the values at of the stochastic process at the ith scenario of the lth data set.

An estimation of PVaR can be obtained using the method described in the previous section with data set A^l, denote it by θ^l. Since θ^l relies on data set A^l, we take it as a function of the data set. That is,

$$\theta^l = F(A^l). \tag{9.19}$$

θ^l estimated this way is taken as a sample of $PVaR_\alpha(x)$, suppose we obtain M samples of $PVaR_\alpha(x)$ with this method on M data sets.

Let the true value of $PVaR_\alpha(x)$ be μ, which is unknown. When the data sets are produced independently, $\theta^1, \cdots, \theta^M$ can be taken as independent samples of a normal distribution with μ as the mean, that is,

$$\theta^l \sim N(\mu, \sigma^2), l = 1, 2, \cdots, M. \tag{9.20}$$

Let the average of the M samples of $PVaR_\alpha(x)$ be $\tilde{\theta}$, it is reasonable to take $\tilde{\theta}$ as an estimation of μ.

$$\tilde{\theta} = \frac{1}{M}\Sigma_{l=1}^{M}\theta^{l}. \tag{9.21}$$

Denote the derivation of the M samples by s^2,

$$s^2 = \frac{1}{M-1}\Sigma_{l=1}^{M}(\theta^{l} - \tilde{\theta})^2, \tag{9.22}$$

we know from statistics that $\frac{\tilde{\theta}-\mu}{\sqrt{\frac{s^2}{M}}}$ follows the t-distribution with $M-1$ degrees of freedom. That is,

$$\frac{\tilde{\theta}-\mu}{\sqrt{\frac{s^2}{M}}} \sim t_{M-1}. \tag{9.23}$$

Hence the $1-\beta$ confidence interval of μ is as follows,

$$\left[\tilde{\theta} - t_{M-1}\left(\frac{\beta}{2}\right)\sqrt{\frac{s^2}{M}}, \tilde{\theta} + t_{M-1}\left(\frac{\beta}{2}\right)\sqrt{\frac{s^2}{M}}\right], \tag{9.24}$$

where $t_{M-1}\left(\frac{\beta}{2}\right)$ is the "$\frac{\beta}{2}$th percentile" of this probability, as shown in Fig. 9.3.

Define

$$B_{1-\beta}(M) = t_{M-1}\left(\frac{\beta}{2}\right)\sqrt{\frac{s^2}{M}} \tag{9.25}$$

Fig. 9.3 t-Distribution and the "$\frac{\beta}{2}$th percentile" of this probability

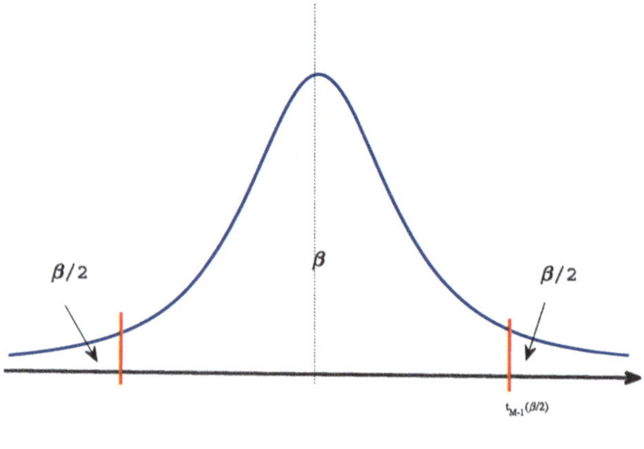

we have the following formula from (9.24),

$$|\tilde{\theta} - \mu| \leq B_{1-\beta}(M). \tag{9.26}$$

Hence $B_{1-\beta}(M)$ is a upper bound for the estimation error with confidence $1-\beta$. In other words, we can state with confidence $1-\beta$ that the largest estimation error is not bigger than $B_{1-\beta}(M)$. The quality of estimation (9.21) can be calculated using formula (9.25) for given β and M; this is one meaningful feature of this method.

On the other hand, we can use formula (9.25) to determine the least number of simulations which is needed to meet the quality requirement for the estimation. Because the more the samples we use in formula (9.21), the closer the estimation will approach to the true value of PVaR, hence estimation error is a decreasing function of sample number M, the upper bound of the estimation $B_{1-\beta}(M)$ is a decreasing function of sample number M.

Using the relation between the upper bound of estimation and number of samples, it is possible to determine the least number for samples in order to meet certain predetermined requirement for the estimation. We will illustrate the method in next section using a numerical example.

9.6 PVaR Estimation Experiments with MC Simulation Method

This section conducts computing experiments for three purposes. The first purpose is to illustrate the method proposed in the previous section in estimating the PVaR of an investment. We take an investment in IBM stock as the example, and estimate the PVaR of the investment using the proposed MC simulation method.

The second purpose is to determine the least number of simulations for getting an estimation of PVaR with a predetermined precision. We will explore the relation between the number of simulations and the upper bound of estimation error, and use this relation to determine the least number of simulations for obtaining a qualified estimation.

The third purpose is to compare the market risk of the investment measured by PVaR with that measured by VaR.

9.6.1 Estimation of PVaR with Monte Carlo Simulation

An investment situation: an investor purchased one share of IBM stock on April 1st, 2019, he will hold the stock for at most 1 year, and he will sell the stock after holding it for 3 months. That is to say, the investor will sell the stock in some time between July 1st, 2019 and April 1st, 2020, the possible liquidation period is 9 months from July 1st, 2019.

We need to make some assumption about the future price of IBM stock to make Monte Carlo simulation to generate scenarios for the prices of IBM stock. Since making market assumption is not the main issue of this paper, we follow a common assumption about stock price, and assume that the future price of IBM stock is a geometric Brownian motion, i.e., stock price at time t can be expressed with the following formula:

$$\omega(t) = \omega(0)exp[(\mu - \sigma^2/2)t + \sigma B(t)], \tag{9.27}$$

where $\omega(0)$ is the stock price at the beginning, μ and σ are parameters, and $B(t)$ is the standard Brownian motion.

The price of IBM stock on April 1st of 2019 is 143.30$, so we set $\omega(0)$ in formula (9.27) to 143.30, and use the following parameters in generation scenarios for the prices of IBM stock in 1 year beginning from April 2nd, 2019.

$$\mu = 0.15, \quad \sigma = 0.20.$$

Firstly, we follow the geometric Brownian motion model (9.27) in creating scenarios for the daily stock price in 1 year beginning from April 2nd, 2019. After generating 100 scenarios for the daily prices in 1 year, we calculate the profit rate and loss rate for each possible holding period, and estimate the PVaR at two confidence levels (95%, 90%) with the method described in Sect. 9.4, estimation is taken as one sample of PVaR.

Secondly, we repeat the above process 100 times, producing 100 samples of PVaR. The 100 samples for $PVaR_{0.95}$ and $PVaR_{0.90}$ are shown in Fig. 9.4. The average of the 100 samples is taken as an estimation of PVaR as shown in Table 9.3.

According to the t-Distribution, we have $t_{M-1}(\frac{\beta}{2}) = t_{99}(0.05) = 1.66$. Hence the upper bound of estimation error with confidence 90%, denoted by $B_{0.90}(100)$, can be obtained using formula (9.25), we fill the results in Table 9.3.

Hence we obtain an estimation of PVaR of the investment, and the upper bound of estimation error as well.

9.6.2 *Least Number of Simulations for Getting a Qualified Estimation of PVaR*

To see the relations between the number of simulations M and the upper bound of estimation error $B_{1-\beta}(M)$, we do estimation experiments using different number of simulations M.

For each M in set $\{100, 200, 300, 400, 500, 600, 700, 800, 900, 1000\}$, we produce M samples of $PVaR_{0.95}$ using the historical simulation method. The distribution of the samples of $PVaR_{0.95}$ in two cases (M = 500 and M = 1000) are shown in

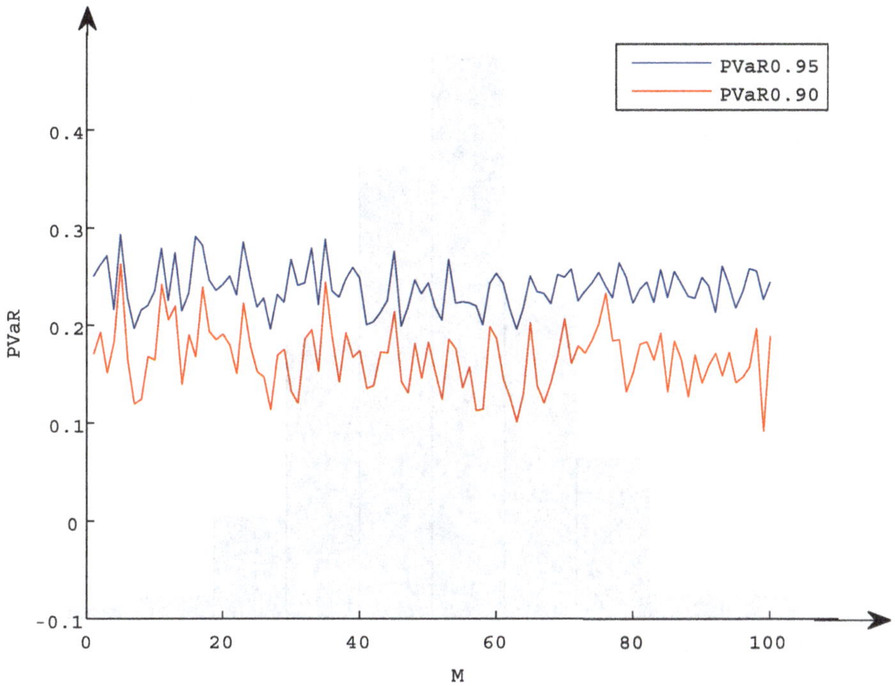

Fig. 9.4 Samples of $PVaR_{0.95}$ and $PVaR_{0.90}$

Table 9.3 Estimation of $PVaR_{0.95}$ and $PVaR_{0.90}$: market risk of an investment on IBM stock

Confidence level	95%	90%
$PVaR_{\alpha}(x)$	23.87%	19.93%
$B_{0.90}(100)$	0.37%	0.32%

Figs. 9.5 and 9.6, respectively. The average of the samples is taken as the estimation $PVaR_{0.95}$, the results are shown in Table 9.4.

According to the t-Distribution, $t_{M-1}(0.05) = 1.66$ for $M = 100$, and $t_{M-1}(0.05) = 1.645$ for $M \geq 200$. For each M in $\{100, 200, 300, 400, 500, 600, 700, 800, 900, 1000\}$, we calculate the upper bound of estimation error with 90% confidence $B_{0.90}(M)$ using formula (9.25), the results are also shown in Table 9.4.

To see the relation between M and upper bound of estimation error more clearly, we plot the upper bounds of estimation error in Fig. 9.7. We see from Fig. 9.7 that the bound of estimation error is a decreasing function of the number of simulations, the more simulations we make, the smaller the upper bound of estimation error will be.

When certain requirement on the estimation error is given, we can determine the least number of simulations under which the estimation error will be smaller than the required. For instance, if estimation error is required to be not bigger than 0.15%, we see from Fig. 9.8 that the least number of simulation is 700. That is to say, the average of 700 samples is an estimation of $PVaR_{0.95}$ with estimation error not bigger than 0.15%.

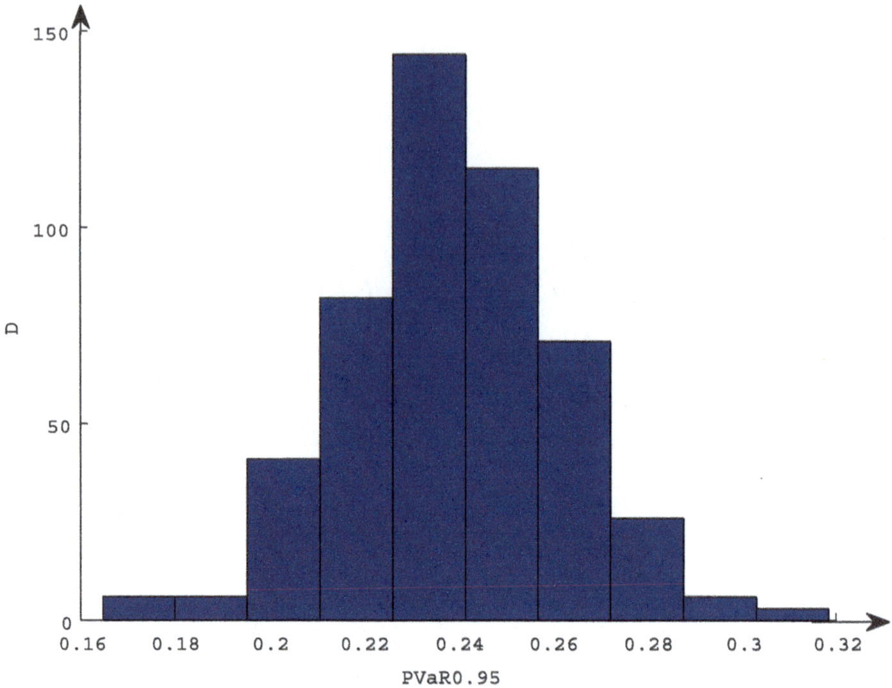

Fig. 9.5 Distribution of 500 samples of $PVaR_{0.95}$

9.6.3 Comparison of Market Risk Measured in PVaR with that in VaR

To compare risk measured by PVaR with risk measured with VaR, we calculate the VaR of holding IBM stock one share for 1 year.

We use the same 100 data sets for the stock prices generated by Monte Carlo simulation in Section 9.6.1, and take the last price in each simulated scenario as the price of IBM stock in the last day of the investment term. We estimate the VaR with the scenario simulation method and obtain 100 samples for VaR. The average of the 100 samples of VaR is taken as its estimation($M = 100$).

The estimation of VaR at two confidence levels together with the corresponding PVaR are summarized in Table 9.5.

We see from the above comparison that risk over a period of time is greater than that at the end of the period. This signifies that PVaR is a proper alternative when risk over a period of time is of concern.

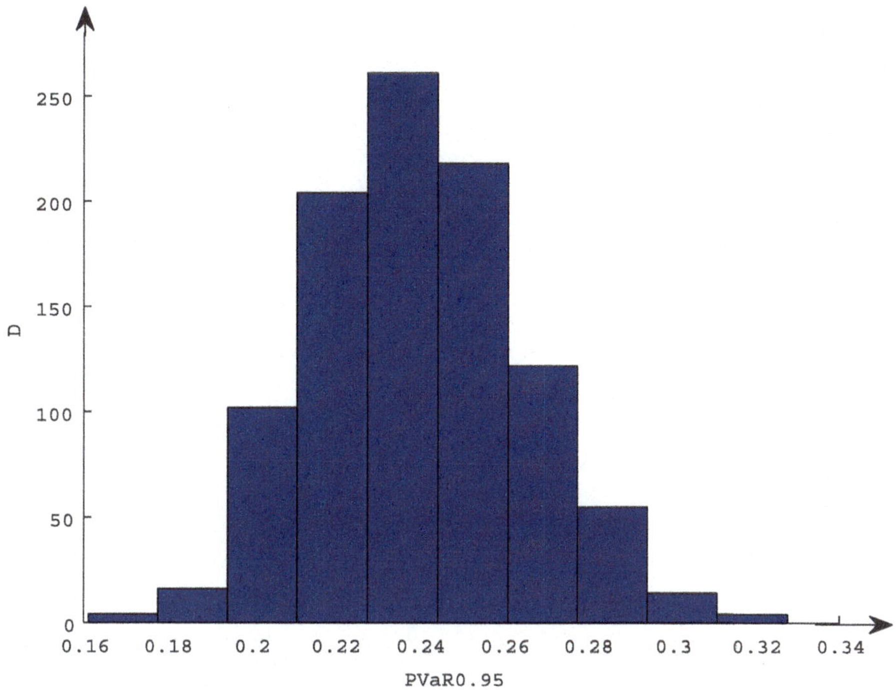

Fig. 9.6 Distribution of 1000 samples of $PVaR_{0.95}$

Table 9.4 Relation between M and Bound of estimation error of $PVaR_{0.95}$

Number(M)	100	200	300	400	500	600	700	800	900	1000
$PVaR_{0.95}$ (%)	23.87	24.22	23.84	24.00	23.91	23.92	23.91	23.95	24.16	23.99
$B_{0.90}(M)$ (%)	0.37	0.28	0.24	0.20	0.17	0.16	0.15	0.14	0.13	0.13

9.7 Conclusions and Remarks

This paper introduced PVaR, a risk indicator proposed for measuring market risk of an investment when the investment term is not fixed, a more general investment situation in reality, and presented several methods for computing PVaR corresponding to different situations.

PVaR can be obtained from solving an equation under some special conditions, but can only be estimated with numerical methods. This paper presented three methods for computing PVaR, one analytical method and two numerical methods.

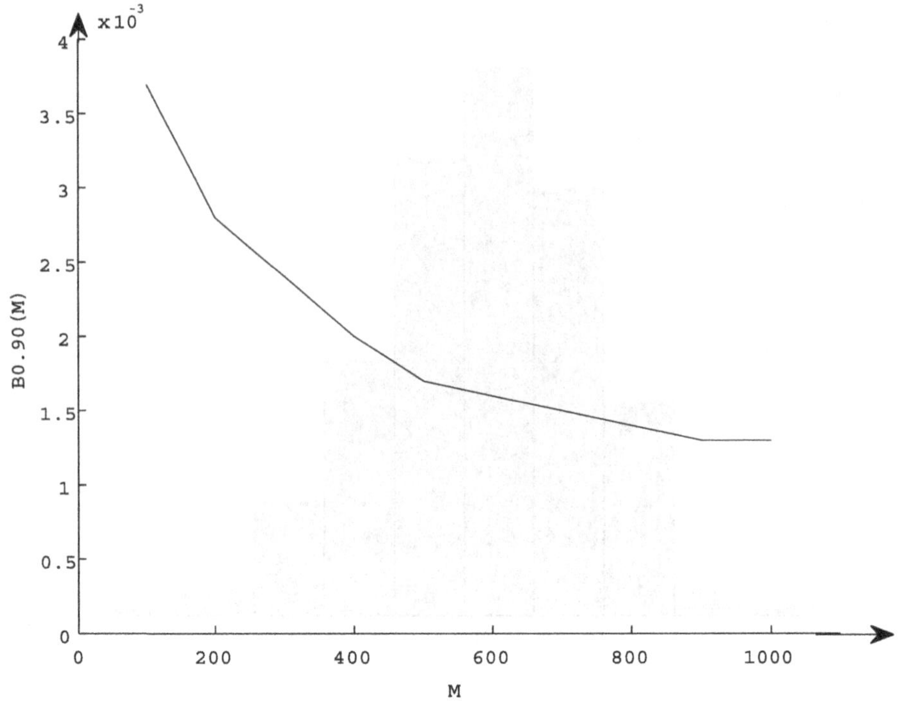

Fig. 9.7 Relation between M and Bound of estimation error of $PVaR_{0.95}$

As all other risk indicators up to date reflect market risk of an investment at a specific future time, PVaR measures the market risk of an investment over a period of time, it is a brand new risk indicator from this point of view. We have been working on investment decision problems using PVaR as the risk measure since this notion was proposed in early 2010s. For instance, we established a PVaR minimization model for investment decision and suggested to solve the model by solving an equivalent model in Huo et al. (2020). Due to the multi-peak property of PVaR, investment decision models with PVaR included turns out to be difficult to solve by conventional optimization methods, one challenging issue we are working on.

I believe traditional variance or VaR-based investment decision theory can be renewed by using PVaR as risk measure and wish more researchers will join the research along this direction.

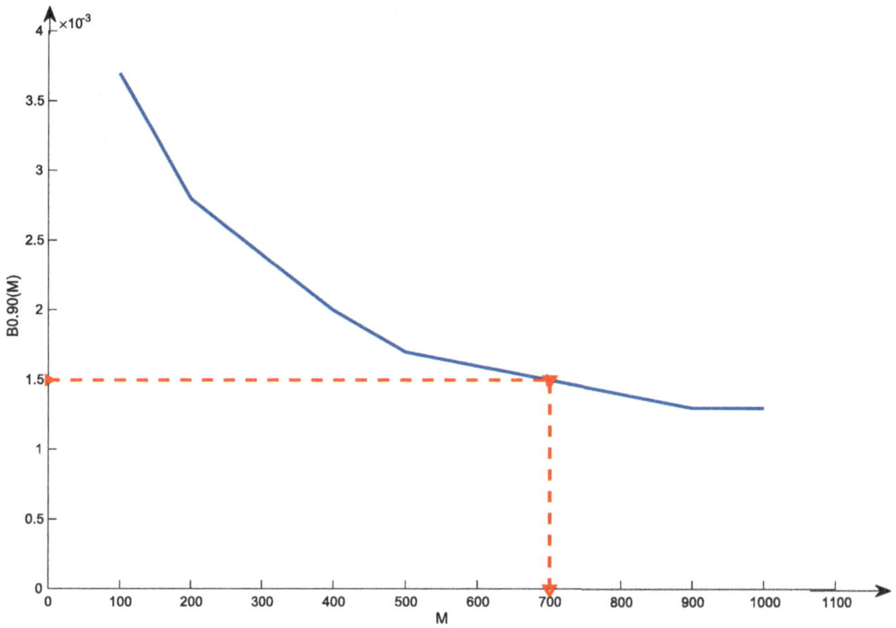

Fig. 9.8 Least number of simulations needed to get an estimation with required quality

Table 9.5 Comparison of PVaR and VaR: market risk of an investment in IBM stock

Confidence level	95%		90%	
Risk indicator	PVaR	VaR	PVaR	VaR
Risk value estimated	23.87 %	16.74%	19.93%	11.35%

Acknowledgments The author acknowledges the support from the Scientific Research (C) grant of Japan Society for the Promotion of Science (19K01757).

References

Acerbi, C., & Tasche, D. (2002). Expected Shortfall: a natural coherent alternative to Value at Risk. *Economic Notes*, *31*(2), 379–388.

Artzner, P., Delbaen, F., Eber, J. M., & Heath, D. (1999). Coherent Measures of Risk. *Mathematical Finance*, *9*, 203–228.

Bawa, V. S., & Lindenberg, E. B. (1977). Capital market equilibrium in a mean-lower partial moment framework. *Journal of Financial Economics*, *5*, 189–200.

Benati, S. (2004). The computation of the worst conditional expectation. *European Journal of Operational Research*, *155*(2), 414–425.

Choudhry, M., & Tanna, K. (2006). *An introduction to value-at-risk*. John Wiley and Sons.

Dowd, K. (2002). *An introduction to market risk measurement*. John Wiley and Sons.

Huo, Y., Xu, C., & Shiina, T. (2020). Modeling and solving portfolio selection problems based on PVaR. *Quantitative Finance*, *20*(12), 1889–1898.

Huo, Y., Xu, C., & Shiina, T. (2021). Period value at risk and its estimation by Monte Carlo simulation. *Applied Economics Letters*, 1–5. https://doi.org/10.1080/13504851.2021.1958136. Published Online: 2021-07-27.

Karatzas, I., & Shreve, S. (1991). *Brownian motion and stochastic calculus* (2nd ed.). Springer.

Konno, H., & Yamazaki, H. (1991). Mean absolute deviation portfolio optimization model and its application to Tokyo Stock Market. *Management Science*, *37*, 517–531.

Markowitz, H. (1952). Portfolio selection. *Journal of Finance*, *7*, 77–91.

RiskMetrics Group. (1996). RiskMetrics - Technical Document. J.P. Morgan/Reuters, NY.

Rockafellar, R. T., & Uryasev, S. (2000). Optimization of conditional value-at-risk. *Journal of Risk*, *2*(3), 21–41.

Tasche, D. (2002). Expected shortfall and beyond. *Journal of Banking and Finance*, *26*(7), 1519–1533.

Xu, C., Huo, Y., & Osaka, K. (2012). Period of value at risk and its estimation by simulation. In *Proceedings of the Asian Conference of Management Science and Applications 2012 (CD-ROM), China, September, 2012* (pp. 48–50).

Young, M. R. (1998). A minimax portfolio selection rule with linear programming solution. *Management Science*, *44*(5), 673–683.

Chapter 10
An Agent-Based Approach to Stability of Complete, Directed, and Signed Social Networks with Loops

Takehiro Inohara

Abstract An agent-based approach to social networks is developed, and some effects of types of agents to systemic properties of the social networks are clarified. Complete, directed, and signed social networks with loops are investigated, and three types of agents, that is, Heider agents, Newcomb agents, and converse Newcomb agents, are treated. The converse Newcomb agents are newly defined in this work and are proved to have strong effects on self-attitudes, symmetry of attitudes, and stability of the social networks. This work deals with three types of stability, that is, Heider's stability, Newcomb's stability, and converse Newcomb stability. The converse Newcomb stability is also newly defined in this work. One of the main propositions establishes the equivalence between Heider's stability and converse Newcomb stability in complete, directed, and signed social networks with loops. This removes redundancy of the traditional definition of Heider's stability and refines so-called "Structure Theorem" on Heider's stability, with respect to complete, directed, and signed social networks with loops.

Keywords Self-attitudes · Symmetry · Balanced graphs · Heider · Newcomb

10.1 Introduction

This work aims to clarify some effects of agents in complete, directed, and signed social networks with loops to systemic properties of the social networks. The systemic properties of social networks treated in this work are self-attitudes, symmetry of attitudes, and stability.

T. Inohara (✉)
Institute for Liberal Arts, Tokyo Institute of Technology, Meguro, Tokyo, Japan

T. Inohara (✉)
Department of Social and Human Sciences, School of Environment and Society, Tokyo Institute of Technology, Meguro, Tokyo, Japan
e-mail: inohara.t.aa@m.titech.ac.jp

K. Kijima et al. (eds.), *Systems Research II*, Translational Systems Sciences 27,
https://doi.org/10.1007/978-981-16-9941-2_10

One of the roots of social network researches is the seminal work on attitudes in psychology by Heider (1946). Cartwright and Harary (1956) developed a signed graph theoretical framework to treat Heider's theory and a rigorous generalization of Heider's concept of stability, called balance, for social networks. Then, a theorem, called "Structure Theorem," is established in Cartwright and Harary (1956), which shows that for all complete, directed, and signed social networks with loops, and Heider's stability is logically equivalent to separability.

Balance of signed social networks has been examined in the literature. For example, random complete signed networks are treated in Deng et al. (2016) with respect to balance of the networks, and relationships between balance and partition of signed graphs are investigated in Figueiredo et al. (2019).

Separability is generalized to clusterability, which is characterized by Davis (1967) with respect to non-existence of semi-cycles with a single arc with a negative attitude. Clusterability under positive self-attitudes is characterized by Inohara (2002a) with respect to Newcomb's stability (Newcomb 1968; Price et al. 1966); that is, it is shown that for all complete, directed, and signed social networks with loops and positive self-attitudes, Newcomb's stability is logically equivalent to clusterability. For complete, directed, and signed social networks with loops in which there may exist negative self-attitudes as well as positive ones, moreover, Newcomb's stability is characterized by Inohara (2002b, 2004a) with respect to general clusterability.

The concepts of stability of social networks are applied to the analysis of reliance on information sources in information networks and information exchange in group decision-making situations. Relationships between stability and structures of reliance in information networks with positive self-attitudes are examined, and Heider's stability and Newcomb's stability are characterized by bisectability and quasi-partitionability, respectively (Inohara 2003a). Information exchange in group decision-making situations such as selecting groups and electing groups is examined by using the concepts of Newcomb's stability, quasi-clusterability (Inohara 2003b), and general quasi-clusterability (Inohara 2004b).

All of these previous researches focus on systemic properties of social networks. In this work, on the other hand, effects of types of agents in social networks to systemic properties of the social networks such as self-attitudes, symmetry of attitudes, and stability are examined.

A type of agents, called converse Newcomb agents, is newly defined in this work. This is a generalized type of Heider agents by definition. As a property of a converse Newcomb agent with respect to self-attitudes, it is shown in this work that a converse Newcomb agent always has a positive attitude toward every negative self-attitude agent. Since a Newcomb agent always has a negative self-attitude toward every negative self-attitude agent (Inohara 2002b, 2004a), this property implies an effect of Heider agents that a complete, directed and signed social network with loops is a positive self-attitude social network if there exists a Heider agent in the social network.

A negative attitude from and a positive attitude to a converse Newcomb agent induce symmetry of negative and positive attitudes between two agents, respectively. These properties imply symmetry of attitudes between two converse Newcomb agents and thus that between two Heider agents, which is a generalization of a result in Cartwright and Harary (1956). The symmetry of attitudes between two Newcomb agents is implied by positive self-attitudes. This generalizes a result in Inohara (2002a).

The existence of a converse Newcomb agent in a complete, directed, and signed social network with loops and Newcomb's stability implies Heider's stability of the social network. This means the equivalence between Heider's stability and Newcomb's stability under the existence of a Heider agent or, more generally, that of a converse Newcomb agent. Converse Newcomb's stability of a complete, directed, and signed social network with loops, moreover, implies Newcomb's stability of the social network. This property of converse Newcomb's stability implies the equivalence between converse Newcomb's stability and Heider's stability in complete, directed, and signed social networks with loops. This removes redundancy in the definition of Heider's stability and refines "Structure Theorem" (Cartwright and Harary 1956), with respect to complete, directed, and signed social networks with loops.

The rest of this work is organized as follows: in the next section, Sect. 10.2, a mathematical framework for treating agents and complete, directed, and signed social networks with loops is constructed. In Sect. 10.3, some propositions on self-attitudes are verified. The symmetry of attitudes is examined in Sect. 10.4, followed by the analysis of effects of types of agents to stability of social networks in Sect. 10.5. In Sect. 10.6, conclusions of this work are provided, and future research topics are discussed.

10.2 Model

In this section, mathematical definitions of *social networks*, *agents*, and *attitudes* are provided. Some mathematical concepts for classification of social networks and agents are also defined. The concepts include *positive* and *negative self-attitude* social networks; *positive* and *negative self-attitude* agents; *Heider's*, *Newcomb's*, and *converse Newcomb's stability* of social networks; and *Heider*, *Newcomb*, and *converse Newcomb* agents.

A *social network* is defined as a set of *agents*, who have relationships, called *attitudes*, among them.

Definition 1 (Social Networks) A *social network* is a pair (N, e) of N and e, where $N = \{1, 2, \ldots, n\}$ is a non-empty and finite set of all *agents* and $e = (e_i)_{i \in N}$ is a list of *attitudes* of agent i, that is, e_i, for each $i \in N$. For each $i \in N$, $e_i = (e_{ij})_{j \in N}$ is a list of agent i's *attitudes* toward agent j, that is, e_{ij}, for each $j \in N$. It is assumed that $e_{ij} \in \{+, -\}$ for each i and $j \in N$. □

For i and $j \in N$, $e_{ij} = +$ and $e_{ij} = -$ mean that agent i has a positive attitude and a negative attitude toward agent j, respectively. A social network is, therefore, a complete, directed, and signed graph with loops. Figure 10.1 is an example of social networks. In this figure, a solid line and a dashed line express that a sign "+" and a sign "−" are assigned on an arc, respectively. All figures in this work are depicted in the same manner.

It should be noted, in particular, that each agent i has an attitude toward her/himself, that is, $e_{ii} = +$ or $e_{ii} = -$. Then, we can define *positive* and *negative self-attitude agents* as follows (see Fig. 10.2):

Definition 2 (Positive and Negative Self-attitude (PSA and NSA) Agents) For $i \in N$, agent i is said to be a *positive* (*negative*, respectively) *self-attitude agent* (a *PSA-agent* (an *NSA-agent*, respectively) for short), if and only if $e_{ii} = +$ ($e_{ii} = -$, respectively). □

If all agents in a social network are PSA-agents (NSA-agents, respectively), then the social network is said to be a *PSA* (an *NSA*, respectively) *social network*.

In the literature, two types of stability are defined for social networks. One is *Heider's stability* (Heider 1946; Cartwright and Harary 1956), and the other is *Newcomb's stability* (Inohara 2002a; Newcomb 1968; Price et al. 1966).

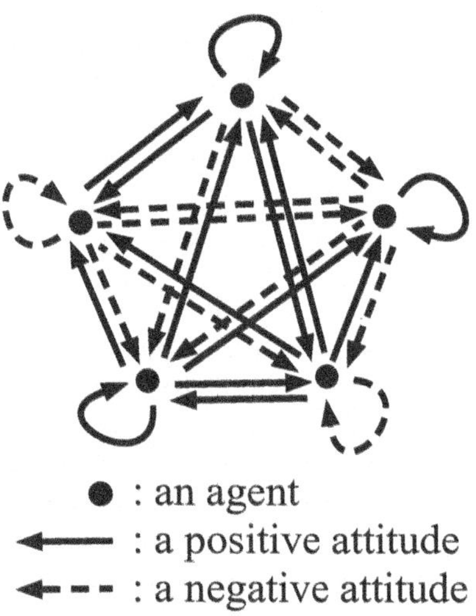

Fig. 10.1 A social network (N, e): A vertex is an agent; A solid line and a dashed line express that a sign "+" and a sign "−" are assigned on an arc, respectively

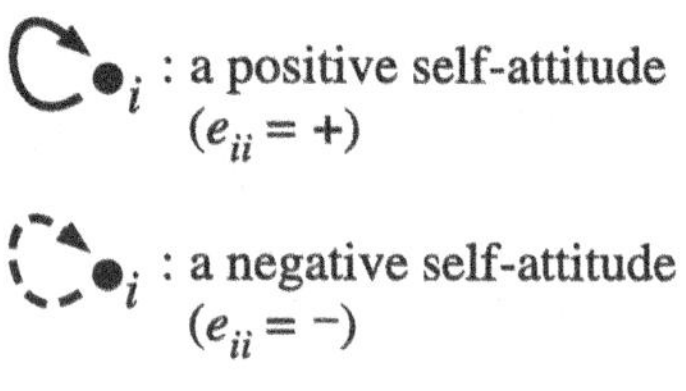

Fig. 10.2 Positive and negative self-attitude agents

Definition 3 (Heider's Stability of Social Networks (Cartwright and Harary 1956; Heider 1946)) A social network (N, e) is said to be *stable in Heider's sense* (*H-stable* for short), if and only if for all i, j, and $k \in N$, it is satisfied that $e_{ij} = +$ if and only if $e_{ik} = e_{jk}$. □

Definition 4 (Newcomb's Stability of Social Networks (Inohara 2002a; Newcomb 1968; Price et al. 1966)) A social network (N, e) is said to be *stable in Newcomb's sense* (*N-stable for short*), if and only if for all i, j, and $k \in N$, it is satisfied that if $e_{ij} = +$, then $e_{ik} = e_{jk}$. □

H-stability requires logical equivalence between $e_{ij} = +$ and $e_{ik} = e_{jk}$, that is, $[e_{ij} = + \Leftrightarrow e_{ik} = e_{jk}]$, for all i, j, and $k \in N$ (see Fig. 10.3). On the other hand, N-stability requires only logical implication from $e_{ij} = +$ to $e_{ik} = e_{jk}$, that is, $[e_{ij} = + \Rightarrow e_{ik} = e_{jk}]$ for all i, j, and $k \in N$ (see Fig. 10.4).

Both of H-stability and N-stability are concepts for social networks. The following two concepts are those for agents on the basis of the ideas in H-stability and N-stability, respectively.

Definition 5 (Heider Agents) An agent $i \in N$ is said to be a *Heider agent* (an *H-agent* for short), if and only if for all j and $k \in N$, it is satisfied that $e_{ij} = +$ if and only if $e_{ik} = e_{jk}$. □

Definition 6 (Newcomb Agents) An agent $i \in N$ is said to be a *Newcomb agent* (an *N-agent* for short), if and only if for all j and $k \in N$, it is satisfied that if $e_{ij} = +$, then $e_{ik} = e_{jk}$. □

The definition of an H-agent $i \in N$ requires logical equivalence between $e_{ij} = +$ and $e_{ik} = e_{jk}$, that is, $[e_{ij} = + \Leftrightarrow e_{ik} = e_{jk}]$ for all j and $k \in N$. Similarly, the definition of an N-agent $i \in N$ requires logical implication from $e_{ij} = +$ to $e_{ik} = e_{jk}$, that is, $[e_{ij} = + \Rightarrow e_{ik} = e_{jk}]$ for all j and $k \in N$. Considering the converse of which the definition of an N-agent requires, we can define a new type of agents, called *converse Newcomb agent* as follows:

Definition 7 (Converse Newcomb Agents) An agent $i \in N$ is said to be a *converse Newcomb agent* (a *cN-agent* for short), if and only if for all j and $k \in N$,

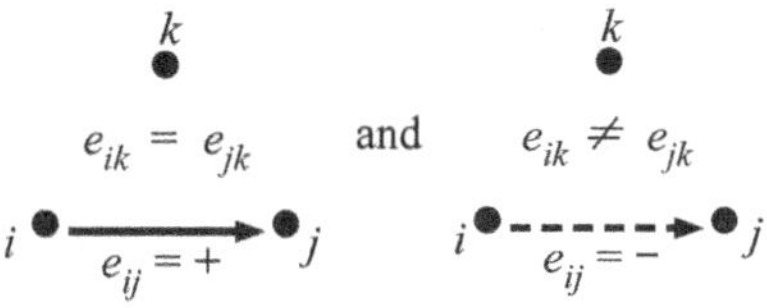

Fig. 10.3 Heider's stability (Cartwright and Harary 1956; Heider 1946)

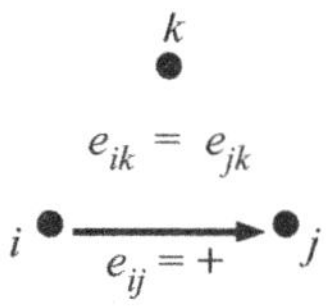

Fig. 10.4 Newcomb's stability (Inohara 2002a; Newcomb 1968; Price et al. 1966)

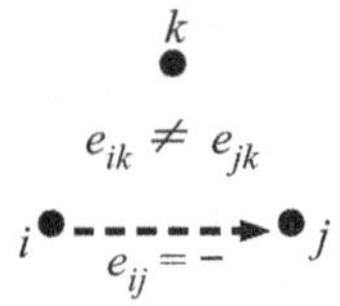

Fig. 10.5 Converse Newcomb's stability

it is satisfied that if $e_{ik} = e_{jk}$, then $e_{ij} = +$, or equivalently, if $e_{ij} = -$, then $e_{ik} \neq e_{jk}$. □

The definition of a cN-agent $i \in N$ requires the logical implication from $e_{ik} = e_{jk}$ to $e_{ij} = +$ (or equivalently, from $e_{ij} = -$ to $e_{ik} \neq e_{jk}$), that is, $[e_{ik} = e_{jk} \Rightarrow e_{ij} = +]$ (or equivalently, $[e_{ij} = - \Rightarrow e_{ik} \neq e_{jk}]$) for all j and $k \in N$, the converse of which the definition of an N-agent does. It is evident from these definitions that an agent is an H-agent if and only if the agent is an N-agent and a cN-agent.

We can naturally define a new stability concept, called converse Newcomb's stability (cN-stability for short), for social networks based on the idea of cN-agents as follows (see Fig. 10.5):

Definition 8 (Converse Newcomb's Stability of Social Networks) A social network (N, e) is said to be *stable in converse Newcomb's sense* (*cN-stable for short*) if and only if for all i, j, and $k \in N$, it is satisfied that if $e_{ik} = e_{jk}$, then $e_{ij} = +$, or equivalently, if $e_{ij} = -$, then $e_{ik} \neq e_{jk}$. □

10.3 Self-attitudes

In this section, some propositions on relationships between self-attitudes and types of agents are verified.

A property of a cN-agent on having a positive self-attitude is verified in the next proposition.

Proposition 1 (cN-Agents and PSA) *A cN-agent is a PSA-agent.* □

Proof Assume that $e_{ii} = -$ for a cN-agent. Then, because agent i is a cN-agent, it is required that $e_{ii} \neq e_{ii}$, which is impossible. Thus, $e_{ii} = +$ for a cN-agent. ■

We immediately have the following corollary from Proposition 1.

Corollary 1 (H-Agents and PSA) *An H-agent is a PSA-agent.* □

Proof An H-agent is a cN-agent by definition. Then, we have the result from Proposition 1. ■

The next example shows that an N-agent may have a positive or a negative self-attitude.

Example 1 (Self-attitudes of N-Agents (Inohara 2002b, 2004a)) Let us consider a social network (N, e), where $N = \{1, 2\}$, as in Fig. 10.6.

Table 10.1 shows that the social network in Fig. 10.6 is N-stable, that is, both agents in the social network are N-agents. In fact, for all combinations of i, j, and $k \in N = \{1, 2\}$, we always have $e_{ik} = e_{jk}$ in the social network in Fig. 10.6. Because agent 1 is an NSA-agent (that is, $e_{11} = -$) and agent 2 is a PSA-agent (that is, $e_{22} = +$), this example shows that an N-agent in a social network may be a PSA-agent or an NSA-agent, which is a different property of an N-agent from that of a cN-agent and that of an H-agent as shown in Proposition 1 and Corollary 1, respectively.

□

The next proposition verifies that an N-agent has a negative attitude toward an NSA-agent.

Proposition 2 (N-Agents and NSA (Inohara 2002b, 2004a)) *An N-agent has a negative attitude toward an NSA-agent.* □

Proof Let $j \in N$ be an N-agent and $i \in N$ be an NSA-agent, that is, $e_{ii} = -$. If $e_{ji} = +$ (see, Fig. 10.7), then it contradicts with the assumption that agent j is an N-agent, because $e_{ji} = +$ and $+ = e_{ji} \neq e_{ii} = -$. Thus, e_{ji} must be $-$. ■

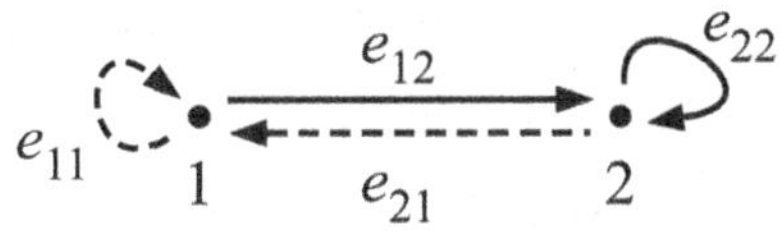

Fig. 10.6 An N-stable social network (N, e)

Table 10.1 N-stability of a social network in Fig. 10.6

i	j	k	e_{ij}	e_{ik}	e_{jk}
1	1	1	−	−	−
1	1	2	−	+	+
1	2	1	+	−	−
1	2	2	+	+	+
2	1	1	−	−	−
2	1	2	−	+	+
2	2	1	+	−	−
2	2	2	+	+	+

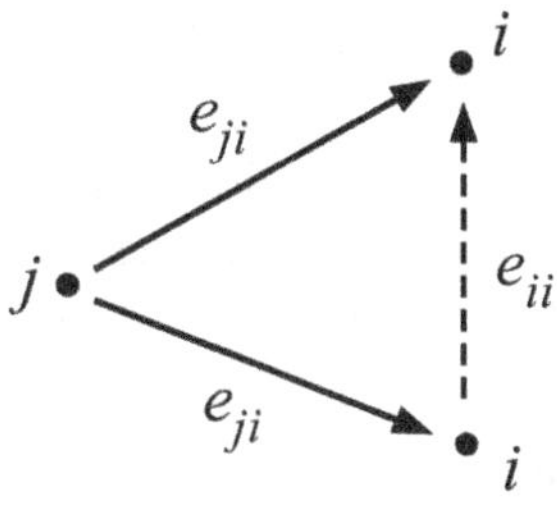

Fig. 10.7 $e_{ji} = +$ contradicts with the assumption that agent j is an N-agent

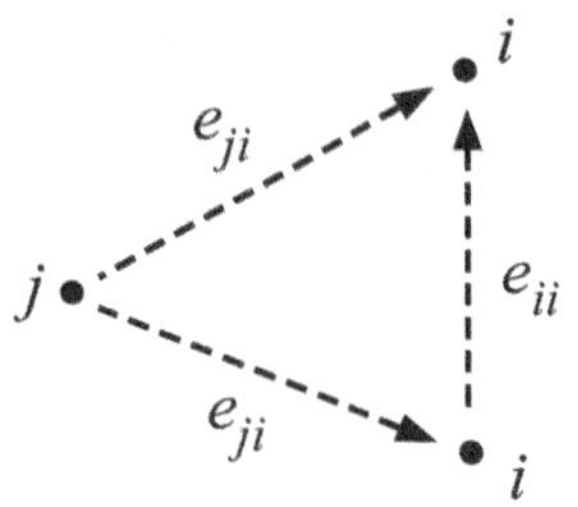

Fig. 10.8 $e_{ji} = -$ contradicts with the assumption that agent j is a cN-agent

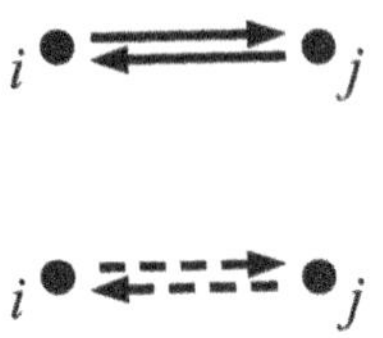

Fig. 10.9 Symmetry of positive and negative attitudes

Proposition 3 (cN-Agents and NSA) *A cN-agent has a positive attitude toward an NSA-agent.* □

Proof Let $j \in N$ be a cN-agent and $i \in N$ be an NSA-agent, that is, $e_{ii} = -$. If $e_{ji} = -$ (see, Fig. 10.8), then it contradicts with the assumption that agent j is a cN-agent, because $e_{ji} = -$ and $- = e_{ji} = e_{ii} = -$. Thus, e_{ji} must be $+$. ■

We immediately have the next corollary from Propositions 2 and 3.

Corollary 2 (H-Agents and PSA Social Networks) *If there exists an H-agent in a social network, then the social network is a PSA social network.* □

Proof Assume that there exists an NSA-agent in the social network. Then, the H-agent has a negative attitude toward the NSA-agent by Proposition 2, because the H-agent is an N-agent by definition. Similarly, because the H-agent is a cN-agent by definition, the H-agent has a positive attitude toward the NSA-agent by Proposition 3. An agent cannot have a positive attitude and a negative attitude toward another agent at the same time, so we have a contradiction. Thus, the NSA-agent cannot exist, which means that the social network is a PSA social network. ■

10.4 Symmetry of Attitudes

In this section, some propositions on relationships between symmetry of attitudes and types of agents are verified.

The symmetry of attitudes of two agents is defined as follows (see Fig. 10.9):

Definition 9 (Symmetry of Agents' Attitudes) For i and $j \in N$, agent i and agent j are said to have *symmetric* attitudes, if and only if $e_{ij} = e_{ji}$. □

Note that for all $i \in N$, agent i and agent i have symmetric attitudes, because we always have $e_{ii} = e_{ii}$. If all pairs of agents in a social network (N, e) have symmetric attitudes, that is, for all i and $j \in N$, $e_{ij} = e_{ji}$, then the social network is said to be *symmetric*.

Two properties of a cN-agent on having symmetric attitudes are verified as follows:

Proposition 4 (A cN-Agent and Symmetry of Negative Attitudes) *For $i \in N$, if agent i is a cN-agent, then it is satisfied that for all $j \in N$, if $e_{ij} = -$, then $e_{ji} = -$.* □

Proof We have that $e_{ii} = +$ by Proposition 1, because a cN-agent is a PSA-agent. $e_{ij} = -$ and $e_{ii} = +$ imply $e_{ji} = -$ by the definition of cN-agents (see Fig. 10.10). ■

Proposition 5 (A cN-Agent and Symmetry of Positive Attitudes) *For $j \in N$, if agent j is a cN-agent, then it is satisfied that for all $i \in N$, if $e_{ij} = +$, then $e_{ji} = +$.* □

Proof We have that $e_{jj} = +$ by Proposition 1, because a cN-agent is a PSA-agent. Assume that $e_{ij} = +$ and $e_{ji} = -$. $e_{ji} = -$ implies $e_{jj} \neq e_{ij}$ by the definition of cN-agents, but this contradicts $+ = e_{jj} = e_{ij} = +$ (see Fig. 10.11). ■

Propositions 4 and 5 immediately imply the next corollary.

Corollary 3 (Symmetry of Attitudes of Two cN-Agents) *For i and $j \in N$, if agent i and agent j are both cN-agents, then they have symmetric attitudes.* □

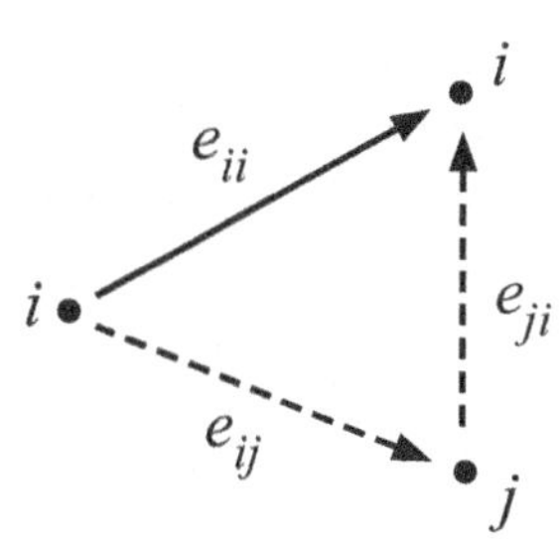

Fig. 10.10 A cN-agent and symmetry of negative attitudes

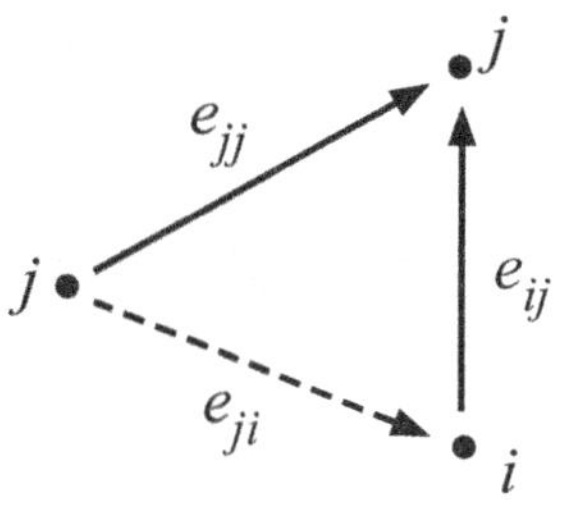

Fig. 10.11 A cN-agent and symmetry of positive attitudes

Proof In the cases that $e_{ij} = +$ and $e_{ij} = -$, we have the result from Propositions 5 and 4, respectively. ■

We also have the following corollary from Corollary 3.

Corollary 4 (Symmetry of Attitudes of Two H-Agents) *For i and $j \in N$, if agent i and agent j are both H-agents, then they have symmetric attitudes.* □

Proof An H-agent is a cN-agent by definition. Then, we have the result from Corollary 3. ■

The next example shows that two N-agents do not always have symmetric attitudes.

Example 2 (Two N-Agents Do Not Always Have Symmetric Attitudes) Let us consider the social network (N, e) in Example 1, that is, $N = \{1, 2\}$, $e_{11} = -$, $e_{12} = +$, $e_{21} = -$, and $e_{22} = +$. As confirmed in Example 1, both agent 1 and agent 2 are N-agents in this case. We see, moreover, that $+ = e_{12} \neq e_{21} = -$, which means agent 1 and agent 2 do not have symmetric attitudes. This example shows that two N-agents do not always have symmetric attitudes. □

We see, in the next proposition, that a positive self-attitude of an N-agent leads to symmetry of positive attitudes.

Proposition 6 (N-Agents with PSA and Symmetry of Positive Attitudes) *For $i \in N$, if agent i is an N-agent and a PSA-agent, then it is satisfied that for all $j \in N$, if $e_{ij} = +$, then $e_{ji} = +$.* □

Proof If $e_{ij} = +$, then we need to have $e_{ii} = e_{ji}$ because agent i is an N-agent. It is satisfied that $e_{ji} = +$, because agent i is a PSA-agent, which implies that $e_{ii} = +$. ■

We immediately have the next corollary on symmetry of attitudes between two N-agents with positive self-attitudes from Proposition 6.

Corollary 5 (Two N-Agents with PSA and Symmetry of Attitudes) *For i and $j \in N$, if agent i and agent j are N-agents and PSA-agents, then they have symmetric attitudes.* □

Proof If both agent i and agent j are N-agents and PSA-agents, then we have, by Proposition 6, that $e_{ij} = +$ if and only if $e_{ji} = +$, which means that agent i and agent j have symmetric attitudes. ■

10.5 Stability of Social Networks

In this section, some propositions on relationships between stability of social networks and types of agents are verified.

We see that an N-stable social network with a cN-agent is H-stable as follows:

Proposition 7 (N-Stable Social Networks with cN-Agents Are H-Stable) *If there exists a cN-agent in an N-stable social network, then the social network is H-stable.* □

Proof Let i_0 be a cN-agent in an N-stable social network (N, e). Then, agent i_0 is an H-agent, because agent i_0 is N-agent, which is implied by the assumption that (N, e) is N-stable. By Corollary 2, we have that for all $i \in N$, agent i is a PSA-agent. By Corollary 5, moreover, (N, e) is symmetric, because all agents in (N, e) are N-agents and PSA-agents. It suffices to show that for all $i, j, k \in N$, if $e_{ij} = -$, then $e_{ik} \neq e_{jk}$. Let us assume that $e_{ij} = -$, and consider two cases, that is, $e_{ik} = +$ and $e_{ik} = -$.

In the case that $e_{ik} = +$ (see Fig. 10.12), we have that $e_{ij} = e_{kj} = -$, because agent i is an N-agent and $e_{ik} = +$. Then, we have that $e_{jk} = -$, because (N, e) is symmetric. This implies that $e_{ik} \neq e_{jk}$.

In the case that $e_{ik} = -$, let us consider two cases, that is, $e_{i_0 i} = +$ and $e_{i_0 i} = -$.

In the case $e_{i_0 i} = +$ (see Fig. 10.13), we have that $e_{i_0 j} = -$ and $e_{i_0 k} = -$, because agent i_0 is an H-agent. Thus, it is satisfied that $e_{jk} = +$, because, again, agent i_0 is an H-agent. This implies that $e_{ik} \neq e_{jk}$.

In the case $e_{i_0 i} = -$ (see Fig. 10.14), we have that $e_{i_0 j} = +$ and $e_{i_0 k} = +$, because agent i_0 is an H-agent. Thus, it is satisfied that $e_{jk} = +$, because, again, agent i_0 is an H-agent. This implies that $e_{ik} \neq e_{jk}$.

In any cases, therefore, we see that $e_{ik} \neq e_{jk}$ if $e_{ij} = -$, which means that (N, e) is H-stable. ■

We immediately have the next corollary on the equivalence between H-stability and N-stability under the existence of an H-agent from Proposition 7.

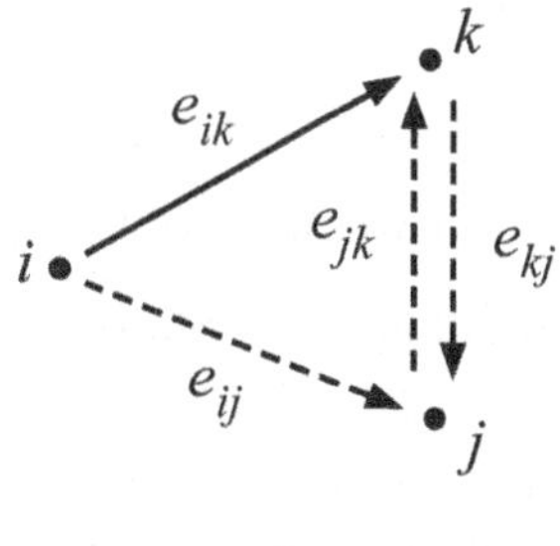

Fig. 10.12 In the case that $e_{ik} = +$

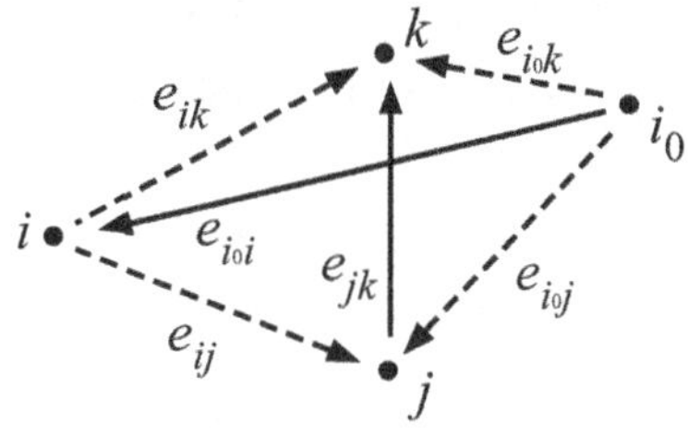

Fig. 10.13 In the case that $e_{ik} = -$ and $e_{i_0 i} = +$

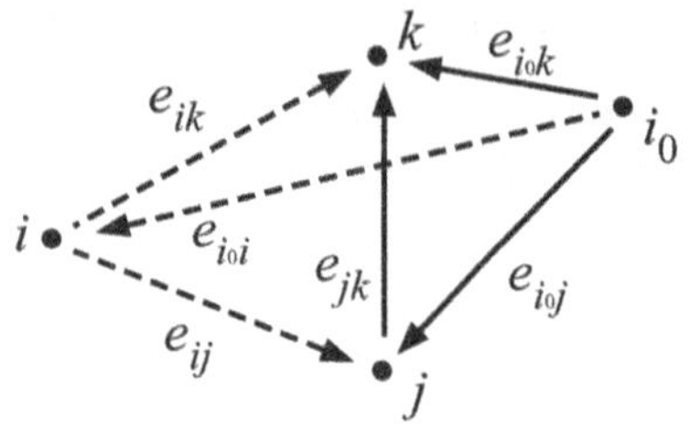

Fig. 10.14 In the case that $e_{ik} = -$ and $e_{i_0i} = -$

Corollary 6 (Equivalence of H-Stability and N-Stability Under the Existence of H-Agents) *If there exists an H-agent (or more generally, a cN-agent) in a social network, then the social network is H-stable if and only if it is N-stable.* □

Proof An H-stable social network is N-stable by definition. If there exists a cN-agent in an N-stable social network, then the social network is H-stable by Proposition 7. Thus, we have the logical equivalence of H-stability and N-stability under the existence of cN-agents. We have the equivalence of H-stability and N-stability under the existence of H-agents, because an H-agent is a cN-agent. ■

The next proposition shows that cN-stability implies N-stability.

Proposition 8 (cN-Stable Social Networks Are N-Stable) *If a social network is cN-stable, then it is N-stable.* □

Proof Consider a cN-stable social network. We have that it is symmetric by Corollary 3. It suffices to show that for all $i, j, k \in N$, if $e_{ij} = +$, then $e_{ik} = e_{jk}$. Let us assume that $e_{ij} = +$, and consider two cases, that is, $e_{ik} = +$ and $e_{ik} = -$.

In the case that $e_{ik} = +$, by the symmetry of the social network, we have that $+ = e_{ji} = e_{ki} = +$. Because agent j is cN-agent, $e_{jk} = -$ contradicts with $e_{ji} = e_{ki}$. We have, therefore, that $e_{jk} = +$, which implies that $e_{ik} = e_{jk}$.

In the case that $e_{ik} = -$, because agent i is cN-agent and $e_{ik} = -$, we have that $+ = e_{ij} \neq e_{kj}$, which implies $e_{kj} = -$. By the symmetry of the social network, we have that $e_{jk} = -$, which implies that $e_{ik} = e_{jk}$. ■

We have two corollaries from Proposition 8.

Corollary 7 (cN-Stable Social Networks Are H-Stable) *If a social network is cN-stable, then it is H-stable.* □

Proof We have, by Proposition 8, that cN-stability implies N-stability. We have the result because cN-stability and N-stability imply H-stability. ■

Corollary 8 (Equivalence of H-Stability and cN-Stability) *A social network is H-stable if and only if it is cN-stable.* □

Proof If a social network is H-stable, then it is cN-stable by definition. We have that cN-stability implies H-stability by Corollary 7. ■

This corollary verifies logical equivalence of H-stability and cN-stability. Because H-stability implies cN-stability by definition, this corollary removes redundancy of the definition of H-stability, that is, we can define an H-stable social network as a cN-stable one.

10.6 Conclusions

The concept of cN-agents was newly defined in this work. Propositions verified in this work showed that cN-agents have strong effects on self-attitudes (Propositions 1 and 3), symmetry of attitudes (Propositions 4 and 5), and stability of social networks (Propositions 7 and 8).

Corollary 8 established equivalence between H-stability and cN-stability. This removes redundancy in the definition of Heider's stability. By using this corollary and the "Structure Theorem" (Cartwright and Harary 1956), moreover, we immediately have equivalence between cN-stability and separability for all complete social networks with loops. It refines the "Structure Theorem" (Cartwright and Harary 1956).

An agent-based approach to social networks was developed in this work. This approach can be applied to the analysis of reliance on information sources in information networks (Inohara 2003a) and information exchange in group decision-making situations (Inohara 2003b, 2004b). Coalition analysis (Inohara and Hipel 2008a,b) and attitude analysis (Inohara et al. 2007) within the framework of the Graph Model for Conflict Resolution (Fang et al. 1993), which deal with multi-participant strategic decision-making situations considering social networks among decision makers, are also fields of application of the agent-based approach to social networks.

This work focused only on complete, directed, and signed social networks with loops. Generalized results for incomplete social networks should be established in the future research opportunities. Small-world social networks (Watts and Strogatz 1998) and scale-free social networks (Barabasi and Albert 1999) are worth being considered as a more general framework.

References

Barabasi, A. L., & Albert, R. (1999). Emergence of scaling in random networks, *Science*, *286*, 509–512.

Cartwright, D. & Harary, F. (1956). Structural balance: a generalization of Heider's theory. *Psychological Review*, *63*, 277–293.

Davis, J. A. (1967). Clustering and structural balance in graphs. *Human Relations*, *20*, 181–187.

Deng, H., Abell, P., Engel, O., Wu, J., & Tan, Y. (2016). The influence of structural balance and homophily/heterophobia on the adjustment of random complete signed networks. *Social Networks*, *44*, 190–201.

Fang, L., Hipel, K. W., & Kilgour, D. M. (1993). *Interactive decision making: The graph model for conflict resolution*. Wiley.
Figueiredo, R., Frota, Y., & Labbe, M. (2019). A branch-and-cut algorithm for the maximum k-balanced subgraph of a signed graph. *Discrete Applied Mathematics*, *261*, 164–185.
Heider, F. (1946). Attitudes and cognitive organization. *Journal of Psychology*, *21*, 107–112.
Inohara, T. (2002a). Characterization of clusterability of signed graph in terms of Newcomb's balance of sentiments. *Applied Mathematics and Computation*, *133*(1), 93–104.
Inohara, T. (2002b). Network stability based on reliability of information sources. In *Proceedings of the 6th Pacific Asia Conference on Information Systems (PACIS2002, Tokyo, Japan, September 2–4, 2002)* (pp. 406–413).
Inohara, T. (2003a). Stability of reliance of information sources and clusterability of information sources. In *Proceedings of the 7th World Multi-Conference on Systemics, Cybernetics and Informatics (SCI 2003, Orlando, Florida, July 27–30, 2003)* (Vol. VII, pp. 225–229).
Inohara, T. (2003b). Clusterability of groups and information exchange in group decision making with approval voting system. *Applied Mathematics and Computation*, *136*(1), 1–15.
Inohara, T. (2004a). Signed graphs with negative self evaluation and clusterability of graphs. *Applied Mathematics and Computation*, *158*(2), 477–487.
Inohara, T. (2004b). Quasi-clusterability of signed graphs with negative self evaluation, *Applied Mathematics and Computation*, *158*(1), 201–215.
Inohara, T., & Hipel, K. W. (2008a). Coalition analysis in the graph model for conflict resolution. *Systems Engineering*, *11*(4), 343–359.
Inohara, T., & Hipel, K. W. (2008b). Interrelationships among noncooperative and coalition stability concepts. *Journal of Systems Science and Systems Engineering*, *17*(1), 1–29.
Inohara, T., Hipel, K. W., & Walker, S. (2007). Conflict analysis approaches for investigating attitudes and misperceptions in the War of 1812. *Journal of Systems Science and Systems Engineering*, *16*(2), 181–201.
Newcomb, T. M. (1968). Interpersonal balance. In R. P. Abelson, E. Aronson, W. J. McGuire, T. M. Newcomb, M. J. Rosenberg, & P. H. Tannenbaum (Eds.), *Theories of cognitive consistency: A sourcebook* (pp. 28–51). Rand-McNally.
Price, K. O., Harburg, E., & Newcomb, T. M. (1966). Psychological balance in situations of negative interpersonal attitudes. *Journal of Personality and Social Psychology*, *3*, 265–270.
Watts D. J., & Strogatz, S. H. (1998). Collective dynamics of 'small-world' networks. *Nature*, *393*, 440–442.

Chapter 11
A Model of Consensus and Consensus Building Within the Framework of Committees with Permissible Ranges of Decision Makers

Takehiro Inohara

Abstract A model of consensus and consensus building is proposed within the framework of voting committees with permissible ranges of decision makers. A group decision-making situation is expressed by a voting committee with the unanimous decision rule, and a negotiation process among decision makers in the situation is expressed as a sequence of decision makers' permissible ranges. Consensus is, moreover, defined as a permissible range of decision makers with a stable alternative and consensus building as a sequence of decision makers' permissible ranges from the status quo to consensus. The existence of consensus and relationships between consensus in a committee, the core of the committee, and Nash equilibrium are investigated.

Keywords Group decision and negotiation · Consensus · Committees · Core · Efficiency · Nash equilibrium

11.1 Introduction

A new model of "consensus" and a definition of "consensus building" are proposed in this work within a framework of voting committees (Peleg 1984; Yamazaki et al. 2000). A committee (Peleg 1984) expresses a group decision-making situation. A negotiation process among decision makers (DMs) in the situation is expressed as a sequence of decision makers' permissible ranges, and "consensus" is defined as a permissible range of decision makers with a "stable alternative (Yamazaki et al.

T. Inohara (✉)
Institute for Liberal Arts, Tokyo Institute of Technology, Meguro, Tokyo, Japan

Department of Social and Human Sciences, School of Environment and Society, Tokyo Institute of Technology, Meguro, Tokyo, Japan
e-mail: inohara.t.aa@m.titech.ac.jp

K. Kijima et al. (eds.), *Systems Research II*, Translational Systems Sciences 27,
https://doi.org/10.1007/978-981-16-9941-2_11

2000)." Then, "consensus building" is defined as a sequence of decision makers' permissible ranges from "status quo" to "consensus."

"Consensus" and "consensus building" are mentioned in the literature as

> Consensus building is a process of seeking unanimous agreement. It involves a good-faith effort to meet the interests of all stakeholders. Consensus has been reached when everyone agrees they can live with whatever is proposed after every effort has been made to meet the interests of all stakeholding parties (page 6 in Susskind (1999)).

and

> A group reaches consensus on a decision when every member can agree to support that decision. Each person may not think it's the very best decision, but he or she can buy into it and actively support its implementation. No one in the group feels that his or her fundamental interests have been compromised. Consensus is not "almost everybody." It's unanimous support for a decision, in the same way that a jury returns a unanimous verdict (page 58 in Straus (2002)).

Since both Susskind (1999) and Straus (2002) deal with the words "agree," "unanimous," and "interests," a model of "consensus" and a definition of "consensus building" should involve these words as keywords. Also, the phrase "live with (Susskind 1999)," which has the meaning "to accept something unpleasant (Oxford Advanced Learner's Dictionary 2000)," is almost the same meaning as the phrase "buy into (Straus 2002)" accompanied with the sentence "[e]ach person may not think it's the very best decision (Straus 2002)." A DM, therefore, should be modeled as an agent who may agree to a decision which is not the best for him/her.

With respect to the difference between "consensus" and "agree," moreover, referring to the sentences "[c]onsensus has been reached when everyone agrees (Susskind 1999)," "A group reaches consensus (Straus 2002)," and "every member can agree (Straus 2002)," the author uses "agree" for describing an individual's state and "consensus" for expressing a group's state. More specifically, in this work, a group is said to reach "consensus" on a decision, if and only if every DM in the group "agrees" to the decision.

In the next section, a framework of voting committees with DMs' permissible ranges is provided on the basis of the framework in Yamazaki et al. (2000). In Sect. 11.3, mathematical definitions of consensus and consensus building are newly proposed, and Sect. 11.4 verifies a relationship between consensus building in a committee and the core of the voting committee. Section 11.5 treats a strategic aspect of consensus building and investigates a relationship between consensus and Nash equilibrium (Nash 1950, 1951). Section 11.6 verifies the existence of consensus in a committee. The last section is devoted to conclusions.

11.2 Preliminaries (Peleg 1984; Yamazaki et al. 2000)

On the basis of the frameworks in Peleg (1984) and Yamazaki et al. (2000), this section gives a framework of voting committees with DMs' permissible ranges. Mathematical definitions of simple games, properness, unanimity, committees, cores of committees, and efficiency are provided in Sect. 11.2.1. Then, in Sect. 11.2.2, definitions of permissible ranges, stable coalitions, and stable alternatives are provided, and two propositions verified in Yamazaki et al. (2000) are introduced.

11.2.1 Committees and Core (Peleg 1984)

A simple game specifies the set of all DMs in a group decision-making situation and the decision-making rule adopted in the situation.

Definition 1 (Simple Games) A simple game is a pair $(N, \mathbb{W})$ of a set N of all DMs and a set $\mathbb{W}$ of all winning coalitions, where (i) $\emptyset \notin \mathbb{W}$ and $N \in \mathbb{W}$ and (ii) if $S \subseteq T \subseteq N$ and $S \in \mathbb{W}$, then $T \in \mathbb{W}$. □

A winning coalition is assumed to have enough power to make the coalition's opinion be the final decision of the group, if every DM in the coalition agrees to the opinion.

Under the properness of a simple game, no two disjoint winning coalitions can be formed at the same time.

Definition 2 (Properness of Simple Games) A simple game $(N, \mathbb{W})$ is said to be proper if and only if for all $S \subset N$, $S \in \mathbb{W}$ implies $N \backslash S \notin \mathbb{W}$. □

A unanimous decision rule, on which this work concentrates, is expressed by a unanimous simple game.

Definition 3 (Unanimous Simple Games) A simple game $(N, \mathbb{W})$ is said to be unanimous if and only if $\mathbb{W} = \{N\}$. □

Evidently, a unanimous simple game is proper.

A group decision-making situation is represented by a committee.

Definition 4 (Committees) A committee C is a 4-tuple $(N, \mathbb{W}, A, (\succsim_i)_{i \in N})$ of a set N of all DMs, a set $\mathbb{W}$ of all winning coalitions, a set A of all alternatives, and a list $(\succsim_i)_{i \in N}$ of preferences $\succsim_i$ on A of DM i for each $i \in N$, where $(N, \mathbb{W})$ is a simple game, $2 \leq |N| < \infty$, and $2 \leq |A| < \infty$. For any $i \in N$, the preference $\succsim_i$ on A of DM i is an element of the set $L(A)$ of all linear orderings on A. □

A relation $\succsim$ on A is said to be a linear ordering on A if and only if $\succsim$ is complete, transitive, and anti-symmetric; that is, (i) for x and y in A, $x \succsim y$ or $y \succsim x$ (complete), (ii) for x, y, and z in A, if $x \succsim y$ and $y \succsim z$, then $x \succsim z$ (transitive), and (iii) for x and y in A, if $x \succsim y$ and $y \succsim x$, then $x = y$ (anti-symmetric). Therefore, $L(A)$ is the set of all complete, transitive, and anti-symmetric relations on A.

When we see a linear ordering $\succsim$ on A as a DM's preference, for x and y in A, $x \succsim y$ means that x is equally or more preferred to y. $x \succ y$ is defined as $x \succsim y$ and $\neg(y \succsim x)$, where $\neg$ denotes "not." If $x \neq y$, then $x \succsim y$ implies $x \succ y$, because $\succsim$ is a linear ordering (in particular, an anti-symmetric relation). For $\succsim \in L(A)$, moreover, $max \succsim$ denotes the most preferred alternative in A in terms of $\succsim$, that is, $max \succsim = a$ if and only if for all $x \in A$, $a \succsim x$. $max \succsim$ is uniquely determined, because $\succsim$ is a linear ordering. Furthermore, for $\succsim \in L(A)$, $\succsim = [x, y, z]$ denotes that x is more preferred to y and y is more preferred to z (and hence x is more preferred to z by the transitivity of r), that is, $x \succ y$ and $y \succ z$ (and hence $x \succ z$), with respect to $\succsim$.

Definition 5 (Cores of Committees) Consider a committee $C = (N, \mathbb{W}, A, (\succsim_i)_{i \in N})$, and the relation Dom on A, which is defined as, for all a and b in A, $a\text{Dom}b$ if and only if there exists $S \in \mathbb{W}$ such that $a \succsim_i b$ for all $i \in S$. For all a and b in A, moreover, $a\text{D}\not{o}\not{m}b$ means "not $a\text{Dom}b$." The core of C, denoted by $Core(C)$, is defined as the set $\{a \in A | \forall b \in A \setminus \{a\}, b\text{D}\not{o}\not{m}a\}$. □

As shown in Appendix, in the case that the simple game $(N, \mathbb{W})$ of the committee $C = (N, \mathbb{W}, A, (\succsim_i)_{i \in N})$ is unanimous, the following three propositions are mutually equivalent: (i) $x \in Core(C)$, (ii) x is Pareto efficient, and (iii) x is strongly Pareto efficient, where Pareto efficiency and strong Pareto efficiency are defined as follows:

Definition 6 (Pareto Efficiency (p. 7 in Osborne and Rubinstein (1994))) Consider a committee $C = (N, \mathbb{W}, A, (\succsim_i)_{i \in N})$. x is said to be Pareto efficient if and only if no $b \in A$ satisfies that $b \succ_i x$ for all $i \in N$. More, x is said to be strongly Pareto efficient if and only if no $b \in A$ satisfies that $b \succsim_i x$ for all $i \in N$ and $b \succ_i x$ for some $i \in N$.

11.2.2 Committees with Permissible Ranges (Yamazaki et al. 2000)

Permissible ranges of DMs allow us to treat the flexibility of the DMs and make it possible to model agents who may agree to a decision which is not the best for them.

Definition 7 (Committees with Permissible Ranges) A committee C with permissible range P, denoted by $C(P)$, is a pair of a committee $C = (N, \mathbb{W}, A, (\succsim_i)_{i \in N})$ and a list $P = (P_i)_{i \in N}$ of permissible ranges P_i of DM i for each $i \in N$, where $\succsim_i \in P_i \subseteq L(A)$ for each $i \in N$. It is assumed that for all $i \in N$ and all x and y in A, if $x \succsim_i y$ and there exists $\succsim \in P_i$ such that $max \succsim = y$, then there exists $\succsim' \in P_i$ such that $max \succsim' = x$. The set of all permissible ranges P_i of DM i is denoted by $\mathbb{P}_i$. □

For $i \in N$ and $a \in A$, DM i is said to have a as one of his/her permissible alternatives if and only if there exists $\succsim \in P_i$ such that $max \succsim = a$. The set of all DM i's permissible alternatives is denoted by $maxP_i$, that is, $maxP_i = \{a \in A | \exists \succsim \in P_i, max \succsim = a\}$. The assumption in Definition 7 can be expressed as follows: if $x \succsim_i y$ and $y \in maxP_i$, then $x \in maxP_i$, which can be regarded as a kind of monotonicity. This assumption reflects the idea that each DM considers his/her interests even when he/she agrees to an alternative which is not the best for him/her.

Let S_a be the set $\{i \in S | \exists \precsim \in P_i, max \precsim = a\}$, that is, S_a denotes the set of all DMs who are members of coalition S and have a as one of their permissible alternatives. Moreover, $\mathbb{W}_{C(P)} = \{S \in W | \exists a \in A, S_a \in \mathbb{W}\}$, that is, $\mathbb{W}_{C(P)}$ denotes the set of all winning coalitions S such that S_a forms a winning coalition for some $a \in A$. In other words, $\mathbb{W}_{C(P)}$ is the set of all winning coalitions which have possibility of cooperation to make their permissible alternatives be chosen as the final decision of the group. An alternative must be permissible for all members in a winning coalition in $\mathbb{W}_{C(P)}$, in order to be the final decision of the group, and such alternatives constitute the set $A_{C(P)}$, that is, $A_{C(P)} = \{a \in A | \exists S \in \mathbb{W}, S_a \in \mathbb{W}\}$, or equivalently, $A_{C(P)} = \{a \in A | N_a \in \mathbb{W}\}$.

There may exist a winning coalition $S \in \mathbb{W}_{C(P)}$ such that all DMs in S have an alternative $a \in A$ as their common permissible alternative, and for each DM $i \in S$, the alternative a is the best for him/her among the alternatives in $A_{C(P)}$. Such a coalition is quite stable in the group decision situation, because each DM in S has no incentives to deviate from the coalition, and there is no need for the DMs in S to invite other DMs into S in order to obtain bigger power. This type of winning coalitions is said to be stable, in this work.

Definition 8 (Stable Coalitions) Consider a committee C with permissible range P, that is, $C(P)$, where $C = (N, \mathbb{W}, A, (\precsim_i)_{i \in N})$, and $\mathbb{W}_{C(P)}$. A winning coalition $S \in \mathbb{W}_{C(P)}$ is said to be stable if and only if there exists $a \in A$ such that (i) $S_a = S$, and (ii) for all $i \in S$ and all $b \in A \backslash \{a\}$, if $b \precsim_i a$, then $b \notin A_{C(P)}$. The set of all stable coalitions in $C(P)$ is denoted by $\overline{\mathbb{W}_{C(P)}}$. □

An alternative that has possibility to be selected as the final choice by some stable coalitions is called a stable alternative, and the set of all stable alternatives, that is, the set

$$\{a \in A \,|\exists S \in \mathbb{W}_{C(P)}, S_a = S \wedge (\forall i \in S, \forall b \in A \backslash \{a\}, b \precsim_i a \rightarrow b \notin A_{C(P)})\},$$

is denoted by $\overline{A_{C(P)}}$.

The next proposition validates that the number of stable alternatives in a committee with a proper simple game is at most one.

Proposition 1 (Coincidence of Final Choice (Yamazaki et al. 2000)) *Consider a committee C with permissible range P, that is, $C(P)$, where $C = (N, \mathbb{W}, A, (\precsim_i)_{i \in N})$, and assume that the simple game $(N, \mathbb{W})$ is proper. Then, the number of elements in $\overline{A_{C(P)}}$ is one, at most.* □

Consider a committee $C = (N, \mathbb{W}, A, (\precsim_i)_{i \in N})$ and an alternative $x \in A$. For $i \in N$, P_i^x denotes the set $\{\precsim \in L(A) \mid (max \precsim) \precsim_i x\}$, which expresses that the DM i's permissible alternatives are those that he/she equally or more prefers to x in terms of DM i's preference $\precsim_i$. P^x denotes the list $(P_i^x)_{i \in N}$ of P_i^x for each $i \in N$, and $C(P^x)$ is a committee with permissible range P^x. In this case, in particular, all DMs have x as one of their permissible alternatives.

The next proposition gives a characterization of the stable alternatives in a committee C with permissible range P^x with respect to the core $Core(C)$ of C.

Proposition 2 (Yamazaki et al. 2000) *Consider a committee* $C = (N, \mathbb{W}, A, (\succsim_i)_{i \in N})$*, and assume that the simple game* $(N, \mathbb{W})$ *is proper. For an alternative* $x \in A$*, it is satisfied that* $\overline{A_{C(P^x)}} = \{x\}$ *if and only if* $x \in Core(C)$*.* □

11.3 Consensus and Consensus Building

This section proposes mathematical definitions of consensus and consensus building.

A negotiation process in a group decision situation is expressed by a sequence of DMs' permissible ranges in a committee. It is assumed in this work that the negotiation process starts from the state, called the status quo, in which each DM agrees only to his/her best alternative. In the process, however, each DM may change his/her permissible range and may come to agree to an alternative which is not the best for him/her.

Definition 9 (Negotiation Processes in Committees) Consider a committee $C = (N, \mathbb{W}, A, (\succsim_i)_{i \in N})$. A negotiation process in C is a sequence $(P^t)_{t \in T}$ of DMs' permissible ranges $P^t = (P_i^t)_{i \in N}$ at time t for each $t \in T$, where $T = \{0, 1, 2, \ldots\}$. $P^0 = (P_i^0)_{i \in N} = (\{\succsim_i\})_{i \in N}$ is called status quo. □

Consensus is defined as a state with a stable alternative in a negotiation process, and a sequence of DMs' permissible ranges from the status quo to the consensus is called consensus building.

Definition 10 (Consensus and Consensus Building) Consider a committee $C = (N, \mathbb{W}, A, (\succsim_i)_{i \in N})$. A negotiation process $(P^t)_{t \in T}$ in C is said to reach consensus at $t^* \in T$ on $x \in A$ if and only if either (i) $t^* = 0$ and $\overline{A_{C(P^0)}} = \{x\}$ or (ii) $t^* > 0$, $\overline{A_{C(P^t)}} = \emptyset$ for all t such that $0 \leq t < t^*$, and $\overline{A_{C(P^{t^*})}} = \{x\}$. In either cases, the sequence $(P^0, P^1, \ldots, P^{t^*})$ is called the consensus building on x in C, and x is said to be consensus through the sequence $(P^0, P^1, \ldots, P^{t^*})$ in C. □

The next example shows that the consensus may change depending on the consensus building process.

Example 1 Consider a committee $C = (N, \mathbb{W}, A, (\succsim_i)_{i \in N})$ such that $N = \{1, 2, 3\}$; $\mathbb{W} = \{\{1, 2, 3\}\}$; $A = a, b, c$; $\succsim_1 = [a, b, c]$; $\succsim_2 = [b, c, a]$; $\succsim_3 = [c, b, a]$. Note that the simple game $(N, \mathbb{W})$ is unanimous. Consider, moreover, the following permission ranges of DMs:

$$P_{11} = \{[a, b, c]\};\ P_{12} = \{[a, b, c], [b, a, c]\};\ P_{13} = \{[a, b, c], [b, c, a], [c, a, b]\};$$

$$P_{21} = \{[b, c, a]\};\ P_{22} = \{[b, c, a], [c, b, a]\};\ P_{23} = \{[b, c, a], [c, a, b], [a, b, c]\};$$

$$P_{31} = \{[c, b, a]\};\ P_{32} = \{[c, b, a], [b, c, a]\};\ P_{33} = \{[c, b, a], [b, a, c], [a, c, b]\}.$$

Then, (i) $P^0 = (P_{11}, P_{21}, P_{31})$, $P^1 = (P_{12}, P_{21}, P_{31})$, $P^2 = (P_{12}, P_{21}, P_{32})$ is a consensus building on $b \in A$ in C; in fact, $\overline{A_{C(P^0)}} = \overline{A_{C(P^1)}} = \emptyset$ and $\overline{A_{C(P^2)}} = \{b\}$; (ii) $P^0 = (P_{11}, P_{21}, P_{31})$, $P^1 = (P_{12}, P_{21}, P_{31})$, $P^2 = (P_{13}, P_{21}, P_{31})$, $P^3 = (P_{13}, P_{22}, P_{31})$ is a consensus building on $c \in A$ in C; and (iii) $P^0 = (P_{11}, P_{21}, P_{31})$, $P^1 = (P_{11}, P_{21}, P_{32})$, $P^2 = (P_{11}, P_{21}, P_{33})$, $P^3 = (P_{11}, P_{22}, P_{33})$, $P^4 = (P_{11}, P_{23}, P_{33})$ is a consensus building on $a \in A$ in C.

11.4 Consensus and Core

The following proposition gives a relationship between consensus building in a committee and the core of the committee.

Proposition 3 *Consider a committee* $C = (N, \mathbb{W}, A, (\succsim_i)_{i \in N})$. *Assume that the simple game* $(N, \mathbb{W})$ *is unanimous. Then, for an alternative* $x \in A$, *there exists a negotiation process* $(P^t)_{t \in T}$ *in* C *which reaches consensus at* $t^* \in T$ *on* $x \in A$ *for some* $t^* \in T$ *if and only if* $x \in Core(C)$. □

Proof Consider a committee $C = (N, \mathbb{W}, A, (\succsim_i)_{i \in N})$, where the simple game $(N, \mathbb{W})$ is unanimous, and an alternative $x \in A$.

First, assume that there exists a negotiation process $(P^t)_{t \in T}$ in C which reaches consensus at $t^* \in T$ on $x \in A$ for some $t^* \in T$. Then, we immediately have from the definition of consensus (Definition 10) that $\overline{A_{C(P^{t^*})}} = \{x\}$. Since the simple game $(N, \mathbb{W})$ is unanimous, that is, $\mathbb{W} = \{N\}$, it is implied that $N_x = N$, that is,

$$\text{for all } i \in N, x \in maxP_i, \tag{11.1}$$

and

$$\text{for all } i \in N \text{ and all } b \in A \backslash \{x\}, \text{ if } b \succsim_i x, \text{ then } b \notin A_{C(P)} \tag{11.2}$$

(see Definition 8).

If $x \notin Core(C)$, then the unanimity of the simple game $(N, \mathbb{W})$ implies that

$$\text{there exists } b \in A \backslash \{x\} \text{ such that for all } i \in N, b \succsim_i x. \tag{11.3}$$

The alternative b in (11.3) satisfies that for all $i \in N$, $b \in maxP_i$, which implies $b \in A_{C(P)}$, by (11.1) and the assumption on DMs' permissible ranges (see Definition 7). Equation (11.3) and $b \in A_{C(P)}$ imply the existence of $b \in A \backslash \{x\}$ such that for all $i \in N$, $b \succsim_i x$, and $b \in A_{C(P)}$, which contradicts (11.2).

Thus, if there exists a negotiation process $(P^t)_{t \in T}$ in C which reaches consensus at $t^* \in T$ on $x \in A$ for some $t^* \in T$, then we have that $x \in Core(C)$.

Second, assume that $x \in Core(C)$. If $\overline{A_{C(P^0)}} = \{x\}$, where $P^0 = (P_i^0)_{i \in N} = (\{\succsim_i\})_{i \in N}$, then the negotiation process $(P^t)_{t \in T}$ reaches consensus on $x \in A$ at $t^* = 0$ (see Definition 10). That is, the sequence (P^0) is the consensus building on x.

If $\overline{A_{C(P^0)}} \neq \{x\}$, then consider the sequence (P^0, P^1), where $P^0 = (P_i^0)_{i \in N} = (\{\succsim_i\})_{i \in N}$ and $P^1 = (P_i^1)_{i \in N} = (P_i^x)_{i \in N}$. Then, we have, by Proposition 2 and the assumption that $x \in Core(C)$, that $\overline{A_{C(P^1)}} = \overline{A_{C(P^x)}} = \{x\}$, which implies that the negotiation process $(P^t)_{t \in T}$ reaches consensus on $x \in A$ at $t^* = 1$.

Thus, if $x \in Core(C)$, then there exists a negotiation process $(P^t)_{t \in T}$ in C which reaches consensus at $t^* \in T$ on $x \in A$ for some $t^* \in T$. ■

This proposition implies that for an alternative in a committee, being a consensus through some sequences is equivalent to be an element of the core of the committee.

11.5 Consensus and Nash Equilibrium

Consider a committee $C = (N, \mathbb{W}, A, (\succsim_i)_{i \in N})$. Then, we can define a game $G_C = (N, (\mathbb{P}_i)_{i \in N}, (\succsim_i')_{i \in N})$ in normal form by defining $\succsim_i'$ for each $i \in N$ based on C as follows: for $P = (P_i)_{i \in N}$ and $P' = (P_i')_{i \in N}$ in $\mathbb{P} = \prod_{i \in N} \mathbb{P}_i$, $P \succsim_i' P'$, if and only if either

- $\overline{A_{C(P)}} = \{a\}$, $\overline{A_{C(P')}} = \{b\}$, and $a \succsim_i b$,
- $\overline{A_{C(P)}} = \overline{A_{C(P')}} = \emptyset$,
- $\overline{A_{C(P)}} = \{a\}$, $\overline{A_{C(P')}} = \emptyset$, and $a \in maxP_i$, or
- $\overline{A_{C(P)}} = \emptyset$, $\overline{A_{C(P')}} = \{a\}$, and $a \notin maxP_i'$.

Among these four conditions, fourth one cannot hold for a committee $C = (N, \mathbb{W}, A, (\succsim_i)_{i \in N})$ such that the simple game $(N, \mathbb{W})$ is unanimous, that is, $\mathbb{W} = \{N\}$, because we always have $a \in maxP_i'$ if $\overline{A_{C(P')}} = \{a\}$. Moreover, if the simple game $(N, \mathbb{W})$ in the committee $C = (N, \mathbb{W}, A, (\succsim_i)_{i \in N})$ is unanimous, then we have the next lemma, which is used in the proof of Proposition 4.

Lemma 1 *Consider $P = (P_i)_{i \in N} = (P_i, P_{-i})$ and $P' = (P_i')_{i \in N} = (P_i', P_{-i}')$ in $\prod_{i \in N} \mathbb{P}_i$, and assume that $P_{-i} = P_{-i}'$. If $maxP_i' \subseteq maxP_i$, then $A_{C(P')} \subseteq A_{C(P)}$.* □

Proof For $x \in A$, consider the sets N_x and N_x', which are defined as $\{i \in N | x \in maxP_i\}$ and $\{i \in N | x \in maxP_i'\}$, respectively. Since it is assumed that $P_{-i} = P_{-i}'$, we have $N_x' \subseteq N_x$ from $maxP_i' \subseteq maxP_i$. Therefore, it is satisfied that if $N_x' = N$, then $N_x = N$, which implies by the unanimity of the simple game $(N, \mathbb{W})$ that $A_{C(P')} = \{a \in A | N'_a = N\} \subseteq \{a \in A | N_a = N\} = A_{C(P)}$. ■

The next proposition shows that if a sequence $(P^0, P^1, \ldots, P^{t^*})$ is consensus building on some alternative $x \in A$ in a committee $C = (N, \mathbb{W}, A, (\succsim_i)_{i \in N})$, then $P^{t^*} = (P_i^{t^*})_{i \in N} \in \prod_{i \in N} \mathbb{P}_i$ is a Nash equilibrium in the game $G_C = (N, (\mathbb{P}_i)_{i \in N}, (\succsim'_i)_{i \in N})$, where $P = (P_i)_{i \in N} \in \prod_{i \in N} \mathbb{P}_i$ is said to be a Nash equilibrium (Nash 1950, 1951) in G, if and only if $(P_i, P_{-i}) \succsim'_i (P'_i, P_{-i})$ for all $i \in N$ and all $P'_i \in \mathbb{P}_i$, where $P_{-i} = (P_1, P_2, \ldots, P_{i-1}, P_{i+1}, \ldots, P_n) \in \prod_{j \neq i} \mathbb{P}_j$.

Proposition 4 *Consider a committee $C = (N, \mathbb{W}, A, (\succsim_i)_{i \in N})$ and the game $G_C = (N, (\mathbb{P}_i)_{i \in N}, (\succsim'_i)_{i \in N})$. Assume that the simple game $(N, \mathbb{W})$ is unanimous. Then, for $P = (P_i)_{i \in N} \in \prod_{i \in N} \mathbb{P}_i$, if $\overline{A_{C(P)}} = \{x\}$ for some $x \in A$ in C, then P is Nash equilibrium in G_C.* □

Proof It suffices to verify $P \succsim'_i P'$ for $i \in N$ and $P' = (P'_i)_{i \in N} = (P'_i, P'_{-i}) \in \prod_{i \in N} \mathbb{P}_i$ such that $P_{-i} = P'_{-i}$. Consider the following three cases: (a) $maxP'_i \subseteq maxP_i$ and $x \in maxP'_i$, (b) $maxP'_i \subseteq maxP_i$ and $x \notin maxP'_i$, and (c) $maxP'_i \supseteq maxP_i$.

(a) Cases $maxP'_i \subseteq maxP_i$ and $x \in maxP'_i$:
First, $x \in maxP'_i$ and $P_{-i} = P'_{-i}$ implies that

$$\text{if } N_x = N \text{ then } N'_x = N, \tag{11.4}$$

where for $x \in A$, N_x and N'_x are defined as the sets $\{j \in N | x \in maxP_j\}$ and $\{j \in N | x \in maxP'_j\}$, respectively.
From $maxP'_i \subseteq maxP_i$ and Lemma 1, we have $A_{C(P')} \subseteq A_{C(P)}$, which implies that

$$\text{if } b \notin A_{C(P)} \text{ then } b \notin A_{C(P')}. \tag{11.5}$$

Then, from the unanimity of the simple game $(N, \mathbb{W})$,

$$\begin{aligned} \overline{A_{C(P)}} = \{x\} &\Rightarrow N_x = N \text{ and } \forall i \in N, \forall b \in A \backslash \{x\}, (b \succsim_i x \to b \notin A_{C(P)}) \\ &\Rightarrow N'_x = N \text{ and } \forall i \in N, \forall b \in A \backslash \{x\}, (b \succsim_i x \to b \notin A_{C(P')}) \\ &\quad \text{(by (11.4) and (11.5))} \\ &\Rightarrow \overline{A_{C(P')}} = \{x\} \quad \text{(by Proposition 1)} \end{aligned}$$

Thus, in this case, $P \succsim'_i P'$ holds by the definition of $\succsim'_i$.

(b) Cases $maxP'_i \subseteq maxP_i$ and $x \notin maxP'_i$:
Since $x \notin maxP'_i$ implies $i \notin N'_x$, we have $x \notin \overline{A_{C(P')}}$.
Assume that $\overline{A_{C(P')}} = \{y\}$ for some $y \in A$ such that $y \neq x$. Then, we have to have $N'_y = N$, where N'_y is defined as the set $\{j \in N | y \in maxP'_j\}$. $N'_y = N$ implies $y \in A_{C(P')}$, and $y \in A_{C(P)}$ follows by $maxP'_i \subseteq maxP_i$ and Lemma 1.

$N'_y = N$ implies $y \in maxP'_i$, too. Then, we have $y \succsim_i x$ under the completeness of $\succsim_i$. In fact, if we do not have $y \succsim_i x$, then we need to have $x \succsim_i y$ by the completeness of $\succsim_i$. By the assumption on DMs' permissible ranges (see Definition 7), $x \succsim_i y$ and $y \in maxP'_i$ imply $x \in maxP'_i$, which contradicts the assumption that $x \notin maxP'_i$.
We see that $y \in A$ satisfies that $y \succsim_i x$ and $y \in A_{C(P)}$, which contradict $\overline{A_{C(P)}} = \{x\}$ and $y \neq x$.
Therefore, $\overline{A_{C(P')}} = \{y\}$ for some $y \in A$ such that $y \neq x$ cannot be satisfied, and then, we have $\overline{A_{C(P')}} = \emptyset$.
Thus, in this case, $P \succsim'_i P'$ holds by the definition of $\succsim'_i$.

(c) Cases $maxP'_i \supseteq maxP_i$:
It suffices to show that $\overline{A_{C(P')}} = \{y\}$ for some $y \in A$ such that $y \neq x$ cannot be satisfied, because this implies from Proposition 1 that either $\overline{A_{C(P')}} = \emptyset$ or $\overline{A_{C(P')}} = \{x\}$, and thus, we have $P \succsim'_i P'$.
Assume that $\overline{A_{C(P')}} = \{y\}$ for some $y \in A$ such that $y \neq x$. If it is satisfied that $x \succsim_i y$ and $x \in A_{C(P')}$, then it contradicts $\overline{A_{C(P')}} = \{y\}$ by the definition of $\overline{A_{C(P')}}$.
If we do not have $x \succsim_i y$, then we need to have $y \succsim_i x$ by the completeness of $\succsim_i$. The assumption $\overline{A_{C(P)}} = \{x\}$ means that $x \in A_{C(P)}$, which implies $x \in maxP_i$. By the assumption on DMs' permissible ranges (see Definition 7), $y \succsim_i x$ and $x \in maxP_i$ imply $y \in maxP_i$.
$\overline{A_{C(P')}} = \{y\}$ implies $y \in A_{C(P')}$, which means $N'_y = N$, where N'_y is defined as the set $\{j \in N | y \in maxP'_j\}$. Since $P_{-i} = P'_{-i}$, we have that $y \in A_{C(P)}$ from $y \in maxP_i$.
$y \succsim_i x$ and $y \in A_{C(P)}$ contradict $x \in \overline{A_{C(P)}}$, and thus, we have $x \succsim_i y$.
From the assumption of $maxP'_i \supseteq maxP_i$ and Lemma 1, we have $A_{C(P')} \supseteq A_{C(P)}$, which implies $x \in A_{C(P')}$, because $x \in A_{C(P)}$ follows $\overline{A_{C(P)}} = \{x\}$.
Consequently, we have both $x \succsim_i y$ and $x \in A_{C(P')}$.

■

By this proposition, we see that a stable alternative, and consequently, consensus (see Definition 10), in a committee is actually stable when we see the committee as a strategic decision situation.

11.6 Existence of Consensus

This section deals with the existence of consensus.

Consider a committee $C = (N, \mathbb{W}, A, (\succsim_i)_{i \in N})$. Then, we can define $max(\succsim_i)_{i \in N}$ as a set $\{x \in A \mid \exists i \in N, x = max \succsim_i\}$ of alternatives. Then, we have the next proposition.

Proposition 5 *Consider a committee* $C = (N, \mathbb{W}, A, (\precsim_i)_{i \in N})$*, and assume that the simple game* $(N, \mathbb{W})$ *of the committee* $C = (N, \mathbb{W}, A, (\precsim_i)_{i \in N})$ *is unanimous. Then, we have that* $\emptyset \neq max(\precsim_i)_{i \in N} \subseteq Core(C)$. □

Proof From the argument in the proof of Proposition 6 in Appendix, it is satisfied, in the setting of this work, that $Core(C) = \{x \in A | \forall b \in A \backslash \{x\}, \exists i \in N, x \succ_i b\}$.

If $x \in max(\precsim_i)_{i \in N}$, then for some $i \in N$, $x = \max \precsim_i$, that is, $\exists i \in N, \forall b \in A \backslash \{x\}, x \succ_i b$, which logically implies that $\forall b \in A \backslash \{x\}, \exists i \in N, x \succ_i b$. Thus, we have $max(\precsim_i)_{i \in N} \subseteq Core(C)$.

We have $max(\precsim_i)_{i \in N}$ is non-empty, because $max \precsim_i$ is uniquely determined for each $i \in N$ from the assumption that $\precsim_i$ is a linear ordering for each $i \in N$. ■

Proposition 5 together with Proposition 3 implies that in a committee $C = (N, \mathbb{W}, A, (\precsim_i)_{i \in N})$ with a unanimous simple game $(N, \mathbb{W})$, there always exists a negotiation process $(P^t)_{t \in T}$ in C which reaches consensus at t^* on x for some $t^* \in T$ and some $x \in A$.

The next example shows that $max(\precsim_i)_{i \in N} = Core(C)$ is not always true.

Example 2 Consider a committee $C = (N, \mathbb{W}, A, (\precsim_i)_{i \in N})$ such that $N = \{1, 2, 3\}; \mathbb{W} = \{\{1, 2, 3\}\}; A = \{a, b, c, d\}; \precsim_1 = [b, a, c, d], \precsim_2 = [c, a, d, b], \precsim_3 = [d, a, b, c]$. Then, we see that $max(\precsim_i)_{i \in N} = \{b, c, d\}$ and $Core(C) = \{a, b, c, d\}$. In fact, $a \in A$ is not dominated by any one of the other alternatives. □

11.7 Conclusions

This work proposed a new model of consensus and a definition of consensus building on the basis of the frameworks in Peleg (1984) and Yamazaki et al. (2000) (Sect. 11.3) and verified some relationships between consensus and core (Proposition 3), between consensus and Nash equilibrium (Proposition 4), and the existence of consensus (Proposition 5). More, Proposition 6 in Appendix indirectly showed a relationship between consensus and efficiency.

Through these propositions, we obtained the following insights on consensus and consensus building:

- For an alternative in committee, being a consensus through some sequences is equivalent to be an element of the core of the committee (Proposition 3).
- Consensus is stable in the sense that it constitutes a Nash equilibrium in the game in normal form, which describes the strategic aspect of the committee (Proposition 4).
- There always exists a negotiation process which reaches consensus (Propositions 3 and 5).
- Consensus is efficient (Propositions 3 and 6).

This work treated stability of consensus as a state in a group decision situation in Sect. 11.5. Instead, we should investigate stability of consensus building as a negotiation process in future research opportunities. This requires modelling a group decision situation as a game in extensive form game (Eichberger 1993; Osborne and Rubinstein 1994) or a graph model within the framework of the Graph Model for Conflict Resolution (Fang et al. 1993; Yasui and Inohara 2007). In order to generalize the existence result in Proposition 5, we need to think of Nakamura's theorem (Nakamura 1979) on the relationship between the non-emptiness of cores of committees and the cardinality of the set of all alternatives. Moreover, strategic information exchange should be involved in the model, and the models by Gibbard (1973) and Satterthwaite (1975) and that by Inohara (2002) may be useful.

Appendix: Core and Efficiency

For $i \in N$, a relation $\succsim_i$ on A is said to be complete, if and only if for x and y in A, $x \succsim_i y$ or $y \succsim_i x$. Also, $\succsim_i$ is said to be anti-symmetric, if and only if for x and y in A, if $x \succsim_i y$ and $y \succsim_i x$ then $x = y$. Note that for x and y in A, $x \succ_i y$ is defined as to satisfiy that $x \succsim_i y$ and $\neg(y \succsim_i x)$, where $\neg$ denotes "not," and that if $x \neq y$ and $\succsim_i$ is anti-symmetric, then $x \succsim_i y$ implies $x \succ_i y$.

Lemma 2 *Assume that $\succsim_i$ is complete. Then, for all b and x in A, we have (i) $\neg(b \succsim_i x)$ if and only if $x \succ_i b$, and (ii) $\neg(b \succ_i x)$ if and only if $x \succsim_i b$.* □

Proof First, assume that $\neg(b \succsim_i x)$. By the completeness of $\succsim_i$, $\neg(b \succsim_i x)$ implies that $x \succsim_i b$. Then, $x \succsim_i b$ together with $\neg(b \succsim_i x)$ means $x \succ_i b$. Second, assume that $x \succ_i b$. By the definition of $\succ_i$, we have that $x \succsim_i b$ and $\neg(b \succsim_i x)$. This implies, in particular, that $\neg(b \succsim_i x)$.

The contraposition of the proposition "$\neg(b \succsim_i x)$ if and only if $x \succ_i b$" is the proposition "$\neg(x \succ_i b)$ if and only if $b \succsim_i x$." Replacing b and x with each other, we have "$\neg(b \succ_i x)$ if and only if $x \succsim_i b$." ■

Lemma 3 *Assume that $\succsim_i$ is complete and anti-symmetric. Then, for b and x in A such that $b \neq x$, we have that $\neg(b \succ_i x)$ if and only if $x \succ_i b$.* □

Proof If $\neg(b \succ_i x)$, then, by the definition of $\succ_i$, we have $\neg(b \succsim_i x \wedge \neg(x \succsim_i b))$, which implies $\neg(b \succsim_i x) \vee x \succsim_i b$. In the case of $\neg(b \succsim_i x)$, by Lemma 2, we have $x \succ_i b$. If $x \succsim_i b$, then we have $x \succ_i b$ by the assumptions of $x \neq b$ and anti-symmetry of $\succsim_i$. Thus, in both cases, we have $x \succ_i b$.

If $x \succ_i b$, then we have, in particular, $x \succsim_i b$, by the definition of $\succ_i$. Then, by Lemma 2, $\neg(b \succ_i x)$. ■

Proposition 6 *Consider a committee $C = (N, \mathbb{W}, A, (\succsim_i)_{i \in N})$, and assume that the simple game $(N, \mathbb{W})$ of the committee $C = (N, \mathbb{W}, A, (\succsim_i)_{i \in N})$ is unanimous. Assume, moreover, that $\succsim_i$ is complete and anti-symmetric for all $i \in N$. Then, for $x \in A$, the following three propositions are mutually equivalent:*

1. *$x \in Core(C)$,*
2. *x is Pareto efficient, and*
3. *x is strongly Pareto efficient.*

Proof By Definition 5, $Core(C)$ is defined as the set $\{x \in A | \forall b \in A \backslash \{x\}, \neg(\exists S \in \mathbb{W}, \forall i \in S, b \succsim_i x)\}$. In the case that the simple game $(N, \mathbb{W})$ of the committee $C = (N, \mathbb{W}, A, (\succsim_i)_{i \in N})$ is unanimous, that is, $\mathbb{W} = \{N\}$,

$$\begin{aligned} Core(C) &= \{x \in A | \forall b \in A \backslash \{x\}, \neg(\forall i \in N, b \succsim_i x)\} \\ &= \{x \in A | \forall b \in A \backslash \{x\}, \exists i \in N, \neg(b \succsim_i x)\}. \end{aligned}$$

We have, moreover, by Lemma 2 and completeness of $\succsim_i$,

$$Core(C) = \{x \in A | \forall b \in A \backslash \{x\}, \exists i \in N, x \succ_i b\}. \tag{11.6}$$

By Definition 6, we have that x is Pareto efficient, if and only if

$$\begin{aligned} \neg(\exists b \in A, \forall i \in N, b \succ_i x) &\iff \forall b \in A, \exists i \in N, \neg(b \succ_i x) \\ &\iff \forall b \in A \backslash \{x\}, \exists i \in N, \neg(b \succ_i x), \end{aligned}$$

since we always have $\neg(x \succ_i x)$ for all $x \in A$ and $i \in N$ from the completeness of $\succsim_i$. We have, moreover, by Lemma 3, and the completeness and anti-symmetry of $\succsim_i$, the above statements are all equivalent to

$$\forall b \in A \backslash \{x\}, \exists i \in N, x \succ_i b. \tag{11.7}$$

Similarly, from Definition 6, x is strongly Pareto efficient, if and only if

$$\begin{aligned} &\neg(\exists b \in A, (\forall i \in N, b \succsim_i x) \land (\exists i \in N, b \succ_i x)) \\ &\iff \forall b \in A, \neg((\forall i \in N, b \succsim_i x) \land (\exists i \in N, b \succ_i x)) \\ &\iff \forall b \in A, (\exists i \in N, \neg(b \succsim_i x)) \lor (\forall i \in N, \neg(b \succ_i x)) \\ &\iff \forall b \in A \backslash \{x\}, (\exists i \in N, \neg(b \succsim_i x)) \lor (\forall i \in N, \neg(b \succ_i x)) \end{aligned}$$

Lemma 2 and the completeness of $\succsim_i$ imply that the above statements are all equivalent to

$$\forall b \in A \backslash \{x\}, (\exists i \in N, x \succ_i b) \lor (\forall i \in N, \neg(b \succ_i x)),$$

and moreover, Lemma 3, and the completeness and anti-symmetry of $\succsim_i$ imply that the previous statement is equivalent to

$$\forall b \in A \backslash \{x\}, (\exists i \in N, x \succ_i b) \lor (\forall i \in N, x \succ_i b),$$

which is equivalent to

$$\forall b \in A \backslash \{x\}, \exists i \in N, x \succ_i b. \tag{11.8}$$

Therefore, from Eqs. (11.6), (11.7), and (11.8), we have the result. ∎

References

Eichberger, J. (1993). *Game theory for economists*. Academic Press, Inc.

Fang, L., Hipel, K. W., & Kilgour, D. M. (1993). *Interactive decision making: The graph model for conflict resolution*. Wiley.

Gibbard, A. (1973). Manipulation of voting schemes: A general result. *Econometrica, 41*, 587–601.

Inohara, T. (2002). Generalizations of the concepts of core of simple games and their characterization in terms of permission of voters. *Applied Mathematics and Computation, 132*(1), 47–62.

Nakamura, K. (1979). The vetoers in a simple game with ordinal preferences. *International Journal of Game Theory, 8*, 55–61.

Nash, J. F. (1950). Equilibrium points in n-person games. *Proceedings of National Academy of Sciences of the USA, 36*, 48–49.

Nash, J. F. (1951). Non-cooperative games. *Annals of Mathematics, 54*(2), 286–295.

Osborne, M. J. & Rubinstein, A. (1994). *A course in game theory*. The MIT Press.

Oxford Advanced Learner's Dictionary. (2000). Oxford University Press.

Peleg, B. (1984). *Game theoretic analysis of voting in committees*. Cambridge University Press.

Satterthwaite M. A. (1975). Strategy-proofness and Arrow's conditions: Existence and correspondence theorems for voting procedures and social welfare functions. *Journal of Economic Theory, 10*, 187–217.

Straus, D. (2002). *How to make collaboration work: Powerful ways to build consensus, solve problems, and make decisions*. Berrett-Koehler Pub.

Susskind, L. E. (1999). *The consensus building handbook: A comprehensive guide to reaching agreement*. California: Sage Publications, Inc.

Yamazaki, A., Inohara, T., & Nakano, B. (2000). New interpretation of the core of simple games in terms of voters' permission. *Applied Mathematics and Computation, 108*(2–3), 115–127.

Yasui, S. & Inohara, T. (2007). A graph model of unanimous decision systems. In *2007 IEEE International Conference on Systems, Man and Cybernetics (Montreal, Canada, October 7–10, 2007)* (pp. 1752–1757).

Chapter 12
A Game Theory Investigation of Contract Between IT Vendor and User in Problems of Information System

Umehara Eiichi

Abstract We model a contract between a user company and an IT vendor as a two-person non-zero-sum game. If a contract does not include compensation for damaged items, the IT vendor does not make any effort to reduce the problems. However, if there is such compensation, depending on a payoff, this game can be a non-implement game, a prisoner's dilemma, or an implement game. No problem reduction efforts are made in the prisoner's dilemma, because this strategy is not Pareto optimum. Therefore, we examine the effectiveness of the incentive contract and find that it does not have any effect. Next, we examine an iterated prisoner's dilemma under a long-term contract and find that the user and the IT vendor both make efforts to reduce the system problems. For this purpose, the user must have a smart retaliation strategy such as TFT.

Keywords IT outsourcing · IT vendor control · IT governance · IT risk management · Game theory

12.1 Introduction

When it comes to IT risks, information system failures can significantly hinder the business of user companies and result in serious damages. For this reason, it is extremely important to establish a failure countermeasure and failure management system for information systems. However, a failure management system cannot be constructed solely by user companies; it needs to be built jointly by the user companies and IT vendors. Therefore, a contract that defines the relationship between a company and an IT vendor is required.

U. Eiichi (✉)
Niigata University of International and Information Studies, Niigata, Japan

Emeritus Professor at Tokyo City University, Tokyo, Japan
e-mail: umehara@nuis.ac.jp

K. Kijima et al. (eds.), *Systems Research II*, Translational Systems Sciences 27,
https://doi.org/10.1007/978-981-16-9941-2_12

In response, companies require to strengthen the management of outsourcers such as IT vendors. Companies specify service levels for contracts with outsourcers in the management system for outsourcing. As for outsourcing management, they must manage outsourcers to prevent accident or fraud, and to additionally have outsourcers take the necessary measures to confirm the service-level agreements stipulated in contracts for outsourcing operations and maintenance.

The development, maintenance, and operation of information systems is often outsourced to specialist IT vendors. Many companies have used application services provided by such specialists, such as shared data centers and ASP services. Therefore, to control system risk, it is important for user companies to establish a management system for outsourced IT vendors.

In light of the above background, we examined a system risk management and analyzed what kind of contract should be made between the user company and the outsourcer (hereinafter, IT vendor) using the framework of game theory from the viewpoint of damages and incentives.

In this chapter, we review system risk management and game theory in Sect. 12.2, and in Sect. 12.3, we model the relationship between a user company and an IT vendor using the framework of game theory. We analyze the effect of damages based on game theory in Sect. 12.4 and that of incentives in Sect. 12.5. In Sect. 12.6, we model the effects a repetitive game. We conclude in Sect. 12.7 with a brief summary and mention of future work.

12.2 Background

12.2.1 IT Risk from Human Resources

In IT risk, a system trouble serious influence on an operation. In particular, human resources are often to blame for system failures, such as black boxing due to the increase in outsourcing and the deterioration of employees' skills in user companies. There are also problems such as the reduction of skilled personnel due to the retirement of veteran employees and cuts to personnel development programs due to company-wide cost reduction.

12.2.2 IT Outsourcing and Compensation for Damages

Management outsourced to third-party IT vendors has been recognized as an important issue for system risk management. Therefore, there is an urgent need for the establishment of a system for vendor management. This led to research on RFP model contracts and the proposal of model contracts for damages and incentives.

However, no specific formulation has been made regarding damages and incentives. Generally speaking, when a disability problem occurs, the damage needs to be resolved after mutual consultation. Therefore, we come up with the idea of using game theory to model damages and incentives in contracts with IT vendors and to provide guidelines for such implementation.

12.2.3 Game Theory

Axelrod (1997) modeled norms as a repeated game and showed that, in order to maintain the norm and prevent free riding on a norm, it was necessary to have a meta-norm of a revenge for breaking a norm and a sanction for breaking a norm of others.

Bendor and Swistak (2001) examined eight theorems as a concept of uniformly stable for conditions under which a norm of mutual cooperation is maintained by using an iterated prisoner's dilemma. They found that the uniform stability strategies were normative and smart (i.e., a collaborative strategy) and retaliatory (if the opponent betrays). As examples of these strategies, they mentioned tit for tat (TFT) and grim-trigger (GT: cooperate at first, and if the opponent betrays, always betray after that).

12.3 Modeling of Outsourcing Contract Based on Game Theory

We modeled an outsourcing contract as a two-player game between a user company and an IT vendor. We formulated the game to take into account players, strategies, the possibility of coalitions among players, the type and timing of information acquired by the players, the effect of information on strategies, common knowledge among players, and payoffs.

1. Players.

   The players are a user company (an outsourcer) and an IT vendor (a business consignee).
2. Strategy.

   The user company chooses whether to attempt to solve an IT problem by itself or not. The IT vendor chooses whether to attempt to solve a problem with the outsourced IT operation by itself or not.
3. Possibility of coalition among players.

   If the user company and the IT vendor work together to solve the problem, there is no strategic option because it is a contractual agreement regarding whether the user company or the IT vendor implements the IT problem solution or not. Therefore, it is not a game formulated in this chapter. Modeling when the user companies and IT vendors collaborate is a topic for future research.

   There are two players in our model. However, user companies may outsource the same operations to two or more IT vendors, and if there are three or more players, there is a possibility that a coalition may form. However, we disregard this possibility here because we assume there is only one outsourcer for the same operation. We leave research with three or more players and the formation of coalitions to future work.

4. The type and timing of information acquired by the players, the effect of information on strategies.

   Troubleshooting information is available to both the user company and the IT vendor. We assume the user company also has the trouble countermeasure information, and there is no information asymmetry between the two companies.

5. Common knowledge among players.

   We assume that all troubleshooting information is disclosed and available to both the user company and the IT vendor.

6. Payoffs.

   The payoffs for the user company and the IT vendor are described in the following section. We express payoffs as costs, so the larger the payoff, the greater the disadvantage.

   We assume the costs user companies pay to IT vendors do not change with or without the existence of a damage compensation contract or incentive contract.

   Any IT problem occurs with a certain probability. The IT vendor is not responsible for some of the outsourced operations, such as hardware problems, and any damage caused by these problems should be borne by the user company. Therefore, it is excluded from the modeling in this chapter. Our model is limited to problems caused by operational mistakes, program malfunctions, etc. that are the responsibility of both the user company and the IT vendor. User companies should always bear the cost of these damages.

   If the user company requires the contract with the IT vendor to include damages, the IT vendor may hedge against the risk of damages. For example, the IT vendors can use risk finance to insure their damages. In this case, the IT vendor may try to add the cost of insurance to the delivery price.

   However, we assume these damages are within the scope of the operations entrusted to the IT vendor. In this case, the problem-solving cost is the quality cost for the contracted operation of the IT vendor, which means the IT vendor needs to pay it. Companies purchase services from IT vendors with the understanding of a pre-determined quality, and due to market competition, the price of service at this quality is typically more or less constant. Therefore, our model assumes that delivery prices of the IT vendor do not change regardless of the cost of troubleshooting efforts. The model when the delivery price changes with respect to the quality cost or the risk finance model is a topic for future research.

   Similarly, by introducing incentive contracts, companies may reduce fixed costs paid to IT vendors, which is a kind of completion bonus. However, we do not cover pay-per-success contracts in this work. The scope of the incentive contracts we analyze is limited to the motivation for user companies to encourage IT vendors to improve the quality of their services. Therefore, we assume that IT vendor prices do not change with or without incentives. The analysis of pay-per-success contracts is a topic for future research.

7. Duration of game.

   The duration of one game is one contract. We model contracts over multiple durations as repeated games. In the case of a contract that is automatically renewed, we consider it as one contract until the automatic renewal kicks in and then model the contract for the next duration as a repeated game. In this chapter, the problem probability (T, S, ΔT, ΔS), the damage (D), the fault recovery cost (E_p, E_e, E), and the incentive (b) are the values of one duration.

12.3.1 Payoffs of a User Company

A strategy of the user company is to implement a countermeasure to an IT problem by itself or leave it to the IT vendor without implementing anything. Let T be the probability of an IT problem occurring. If the user company implements troubleshooting measures, the probability of a problem occurrence can be reduced by $\Delta S > 0$. Let E_p be the problem countermeasure cost and let D be the amount of damage paid by the user company in the event of a problem.

12.3.2 Payoffs of an IT Vendor

If the IT vendor makes an effort to reduce the problems, the probability of a problem occurrence will be reduced by $\Delta T > 0$. However, an IT vendor will pay a cost E_e if it makes an effort to reduce the problem.

12.3.3 Claims for Damage from a User Company to an IT Vendor

A contract between a user company and an IT vendor may state that "the user company can charge the IT vendor for part or all of the damage caused by the problem." In this case, the user company does that.

12.3.4 Incentives from a User Company to an IT Vendor

As described in the SLA introduction guidelines for government procurement, an incentive may be included in their contract. This is to encourage IT vendors take voluntary problem prevention measures. If no problem occurs, a user company pays

Table 12.1 Payoff for contract without a claim for damage

		IT vendor	
		Effort	No effort
User company	Effort	$\{-D(T - \Delta S - \Delta T)$ $- b(1 - T + \Delta S + \Delta T) - E_p,$ $- E_e + b(1 - T + \Delta S + \Delta T)\}$	$\{-D(T - \Delta S) - b(1 - T + \Delta S)$ $- E_p,\ b(1 - T + \Delta S)\}$
	No effort	$\{-D(T - \Delta T) - b(1 - T + \Delta T),$ $- E_e + b(1 - T + \Delta T)\}$	$\{-DT - b(1 - T),\ b(1 - T)\}$

Table 12.2 Payoff for contract without a claim for damage and no incentive

		IT vendor	
		Effort	No effort
User company	Effort	$\{2D\Delta T - E, -E\}$	$\{D\Delta T, -E\}$
	No effort	$\{D\Delta T, -E\}$	$\{0, 0\}$

an incentive (b) to an IT vendor. However, in this paper we do not cover pay-per-success contracts with large incentives. This remains a topic for future research.

With the above discussion in mind, Table 12.1 show the payoffs for a case where a user company does not claim damages from an IT vendor.

12.3.5 Payoff in a Case Without a Claim for Damage and No Incentive

Consider the simplest contract with no claims or incentives. In this case, a user company bears all damages of problems that occur. This is the case of $b = 0$ in the payoff in Table 12.1. In game theory, the order of the four payoffs is preserved even if the same value is added to each. Therefore, DT is added to the payoff of a user company, which we denote as $\Delta T = \Delta S$ for simplicity. In addition, if both companies make efforts, the problem reduction rate will be doubled. Furthermore, we assume that the cost of making efforts to reduce problems is the same for both the user company and the IT vendor; that is, let $E_p = E_e = E$. This payoff is shown in Table 12.2.

We found that the IT vendor always takes a {no effort} strategy in this game. Under the contract, the IT vendor does not make any effort to reduce problems. On the other hand, the user company selects {effort} when the reduced damage amount is larger than the effort cost, and {no effort} when the reduced damage amount is smaller than the effort cost.

However, this is not a desirable condition for the user company, since the user company may be able to improve its payoffs if the IT vendor makes efforts to reduce the problems as well. Therefore, the user company needs to consider measures to encourage the IT vendors to make efforts to reduce problems. We consider the following three conditions and analyze their effects.

1. Introduce a claim for damage.
2. Introduce an incentive.
3. Repeated game.

12.4 Introduce a Claim for Damage

12.4.1 Contract for IT Vendor to Pay Full Damage

Consider a case where an IT vendor pays damages. Some contracts between a user company and an IT vendor may include provisions regarding damages. In general, there are contracts that pay damages up to the contract amount or a fixed amount. With a full-payment contract, an IT vendor has a risk in that the amounts of damages may exceed the contract amount. Normally, an IT vendor will not sign such a contract, as the risk is too great. However, let us first analyze a contract where the IT vendor pays in full, which is ideal for the user company.

The payoff in this case is shown in Table 12.3. This is a transposed matrix of Table 12.2, where the user company paid the full amount of damage. In this case, it turns out that the user company can throw problems over the wall to the IT vendor without making any effort to deal with problems.

In practice, there are few contracts between companies and IT vendors where the IT vendor pays the full amount. Therefore, we next consider a case where the IT vendor pays $p(0 \leq p \leq 1)$ of the damage amount and the company pays the rest. As shown in the appendix, the player's strategy in this case depends only on the magnitude relationship between the amount of damage ($D\Delta T$) and the cost of effort (E). As a result, there are three types of games that can be played: a game in which a problem countermeasure has not been implemented, a prisoner's dilemma, and a game in which a problem countermeasure has been implemented. We refer to these as a non-implement game, a prisoner's dilemma, and an implement game, respectively. Table 12.2 depicts the case of $p = 0$, and Table 12.3 depicts the case of $p = 1$.

Table 12.3 Payoff for contract where an IT vendor pays full damage

		IT vendor	
		Effort	No effort
User company	Effort	$\{-E, 2D\Delta T - E\}$	$\{-E, D\Delta T\}$
	No effort	$\{0, D\Delta T - E\}$	$\{0, 0\}$

Table 12.4 Contract for both companies to pay the damage. (() Is the order)

(Game 1) A non-implement game

		IT vendor	
		Effort	No effort
User company	Effort	$\{D\Delta T - E\ (3),\ D\Delta T - E\ (3)\}$	$\left\{D\frac{\Delta T}{2} - E\ (4),\ D\frac{\Delta T}{2}\ (1)\right\}$
	No effort	$\left\{D\frac{\Delta T}{2}\ (1),\ D\frac{\Delta T}{2} - E\ (4)\right\}$	$\{0\ (2),\ 0\ (2)\}$

(Game 2) A prisoner's dilemma

		IT vendor	
		Effort	No effort
User company	Effort	$\{D\Delta T - E\ (2),\ D\Delta T - E\ (2)\}$	$\left\{D\frac{\Delta T}{2} - E\ (4),\ D\frac{\Delta T}{2}\ (1)\right\}$
	No effort	$\left\{D\frac{\Delta T}{2}\ (1),\ D\frac{\Delta T}{2} - E\ (4)\right\}$	$\{0\ (3),\ 0\ (3)\}$

(Game 3) An implement game

		IT vendor	
		Effort	No effort
User company	Effort	$\{D\Delta T - E\ (1),\ D\Delta T - E\ (1)\}$	$\left\{D\frac{\Delta T}{2} - E\ (4),\ D\frac{\Delta T}{2}\ (2)\right\}$
	No effort	$\left\{D\frac{\Delta T}{2}\ (2),\ D\frac{\Delta T}{2} - E\ (4)\right\}$	$\{0\ (3),\ 0\ (3)\}$

12.4.2 Contract for Both Companies to Pay the Damage

The change in the game is the magnitude relationship between the amount of damage and the cost of effort. Therefore, for the sake of simplicity, we consider changes of the game in the case of a contract where both companies pay half the damage ($p = 0.5$). Table 12.4 shows the payoffs for the three types of games in this case.

(Game 1: A Non-Implement Game)

$$D\Delta T \leq E$$

In this game, the troubleshooting effort cost is greater than the damage amount. The Nash equilibrium is {no effort, no effort}. The cost of minor problem countermeasures may be cheaper if measures are taken each time a problem occurs rather than implementing countermeasures in advance. In fact, user companies are unlikely to require IT vendors to pay damages for minor problems. In this case, it is cheaper to not take any measures at all.

(Game 2: A Prisoner's Dilemma)

$$D\frac{\Delta T}{2} < E < D\Delta T$$

In this game, the Nash equilibrium is {no effort, no effort}. An IT vendor makes no effort to reduce the problem. However, if a user company and an IT vendor work together to reduce problems, a Pareto improvement can be achieved.

(Game 3: An Implement Game)

$$E \le D\frac{\Delta T}{2}$$

In this game, there is a big problem when the amounts of damages are more than double the effort. If both the user company and the IT vendor make an effort, the gain can be larger than the amount of damage caused if neither company made an effort. Therefore, both the user company and the IT vendor are motivated to make efforts to reduce problems. When a claim for damage is added to the contract, depending on the magnitude relationship between the damage amount and the problem countermeasure cost, the game of the user company and the IT vendor changes from (1) a non-implement game, (2) a prisoner's dilemma, or (3) an implement game.

For a non-implement game and an implement game where the Nash equilibrium is Pareto optimum, no other problem reduction measures are required. Therefore, we focus on a prisoner's dilemma and consider measures to achieve Pareto improvement by introducing an incentive or a repeated game within the framework of game theory.

12.5 Effect of Incentives

We consider contracts that allow user companies to give incentives to IT vendors if no problem occurs. However, the incentive we analyze in this paper excludes pay-per-success contracts with large incentives, which have a significant impact on the contract amount of IT vendors. In other words, we consider a case where the incentive is less than the amount of damage ($D > b$). For simplicity, we subtracted $(1 - T)b$ from all payoffs, as shown in Table 12.5, and then compared the payoffs in Tables 12.3 and 12.5. We found that, since the user company does not give the IT vendor an incentive to cover the damage amount, the payoff ranking does not change. We conclude that incentives have no effect in this game.

Table 12.5 Payoff for contract with incentive and without a claim for damage ($D\Delta T > E$)

		IT vendor	
		Effort	No effort
User company	Effort	$\{2(D - b)\Delta T - E, \ -E + 2b\Delta T\}$	$\{(D - b)\Delta T - E, \ b\Delta T\}$
	No effort	$\{(D - b)\Delta T, \ -E + b\Delta T\}$	$\{0, \ 0\}$

Table 12.6 Payoff for contract with incentive and a claim for damage

		IT vendor	
		Effort	No effort
User company	Effort	$\{(D-2b)\Delta T - E,\ (D+2b)\Delta T - E\}$	$\left\{(D-2b)\frac{\Delta T}{2} - E,\ (D+2b)\frac{\Delta T}{2}\right\}$
	No effort	$\left\{(D-2b)\frac{\Delta T}{2},\ (D+2b)\frac{\Delta T}{2} - E\right\}$	$\{0,\ 0\}$

Next, Table 12.6 shows cases where the same incentives as in Table 12.4 are given. Again, in order to change the game from a prisoner's dilemma to an implement game, the amount of damage must be $(D + 2b)\Delta T/2 > E$. That is, it must be $b\Delta T \geq E - D\Delta T/2$. However, since the prisoner's dilemma is $D\frac{\Delta T}{2} < E < D\Delta T$, this always holds for non-negative b. From the above analysis, we found that incentives for IT vendors do not affect either strategy. We conclude that incentives have no effect on changing a prisoner's dilemma to an implement game.

12.6 Repeated Game

We found that, while a contract with a claim for damage could create a prisoner's dilemma, a contract with an incentive could not. However, the prisoner's dilemma is still not a desirable situation for a user company.

Therefore, in order to change from the prisoner's dilemma to an implement game, we consider a case where a user company continuously renews the contract with the same IT vendor. Contracts between a user company and an IT vendor are rarely one-time contracts; rather, outsourcing contracts are often multi-year contracts of about 4 years. Many IT vendors are also aiming to receive repeat orders. In other words, a user company and an IT vendor are playing a repeated game. The Nash equilibrium of the one-time prisoner's dilemma is {no effort, no effort}, as stated earlier, but that of the repeated prisoner's dilemma may change to {effort, effort}.

When a user company recognizes that an IT vendor has not make an effort, it can consider taking a smart and retaliatory strategy such as terminating the contract.

In the game in Table 12.4, the payoff of the repeated prisoner's dilemma game when the IT vendor makes an effort n times in a row and does not make an effort for the $n + 1$th time is shown by

$$V\,(\text{user company}) = (n+1)\,(D\Delta T - E) - \frac{D\Delta T}{2}, \tag{12.1}$$

$$V\,(\text{IT vendor}) = n\,(D\Delta T - E) + \frac{D\Delta T}{2}. \tag{12.2}$$

Because the condition of a prisoner's dilemma is $D\frac{\Delta T}{2} < E < D\Delta T$, an IT vendor's payoff is a monotonically increasing function with respect to n. Similarly, the payoff of the user company also increases monotonically with respect to n. Therefore, in the contract between a user company and an IT vendor, a long-term contract can motivate an IT vendor to make problem reduction efforts.

However, if a user company throws it over a wall to an IT vendor without making any effort to address problems, the result is the opposite. In this case, the user company leaves it to the IT vendor and makes no effort. On the other hand, the IT vendor makes an effort n times in a row and does not make an effort for the $(n + 1)^{\text{th}}$ time. The strategy of the user company for the nth time is {no effort, effort} + $(n + 1)$ and thereafter {no effort, no effort}. In this case, the payoffs of the user company and the IT vendor are

$$V\ (\text{user company}) = \frac{nD\Delta T}{2} \tag{12.3}$$

and

$$V\ (\text{IT vendor}) = n\left(\frac{D\Delta T}{2} - E\right), \tag{12.4}$$

respectively.

Equation (12.4) is monotonically decreasing with respect to n, so the optimal number of IT vendor efforts n is 0. In other words, if a user company leaves the problem reduction to the IT vendor and does not itself make any effort, the IT vendor will not make this effort either. From our analysis of this game, we conclude that it is effective for a user company to adopt the following measures in contracts with an IT vendor in order to reduce problems.

1. A user company should set up a claim for damage clause in its contract with an IT vendor. This should change the game of relationships with IT vendors, at least for the prisoner's dilemma.
2. In this prisoner's dilemma, a user company should sign a long-term contract with an IT vendor. This should create a state in which the user company plays a repeated prisoner's dilemma.
3. In this repetitive prisoner's dilemma, a user company should adopt a smart and retaliatory strategy such as TFT or GT. In other words, in order to achieve a mutual cooperative strategy in the repeated prisoner's dilemma, a user company's strategy should be normative and smart (that is, it should initially adopt a cooperation strategy) and later retaliatory (if the other betrays, take a retaliation strategy).

12.7 Conclusion

We modeled a contract between a user company and an IT vendor regarding information system problems as a non-zero-sum two-player game. Our analysis has shown that IT vendors make no effort to reduce problems in the absence of a claim for damage clause. However, we found that the game can be a prisoner's dilemma if a claim for damage clause is introduced. However, even with the prisoner's dilemma, an IT vendor makes no effort to reduce problems. We therefore examined the effectiveness of incentives as a way to get an IT vendor to make efforts to reduce problems, but found that incentives had the effect of changing the prisoner's dilemma.

In general, contracts between a user company and an IT vendor are often long-term rather than one-time. Therefore, we modeled a long-term contract as a repeated prisoner's dilemma. In the repeated prisoner's dilemma, both players can take a cooperative strategy, and similarly, for a game between a user company and an IT vendor, a continuously renewed contract with the same IT vendor allow both players to adopt a collaborative strategy. As a result, the IT vendor might make efforts to reduce the problems. However, this requires the user company to adopt strict betrayal strategies, such as tit for tat (TFT).

When a user company throws a problem reduction effort over a wall to an IT vendor, even in a repeated game, the IT vendor does not typically make voluntary efforts to reduce problems. As the Financial Services Agency pointed out, if a user company wants an IT vendor to take countermeasures against problems, it needs to constantly monitor the IT vendor and impose disciplinary action (e.g., contract termination) if the IT vendor does not make sufficient efforts. In short, the user company needs to take a normative, smart, and retaliatory strategy against an IT vendor in order to enhance IT governance.

As a future research topic, we intend to clarify the criteria for what kind of descriptions should be made for a damage claim in contracts with an IT vendor. We also want to examine incentives in more detail, given that they had no effect on changing a prisoner's dilemma in our game framework. It is interesting to note that some IT vendors have a scheme in which the results of the incentive are reflected in the internal personnel evaluation of the department in charge. In this case, by adding incentives to the contract, it may be possible to encourage the department in charge to make more efforts to reduce problems. This point will be the subject of future research. In addition, when problems occur, not only financial damage but also reputation problems may result for both the user company and the IT vendor. The reputation risk is also an issue for future research.

Appendix: 3 Games (Non-Implement Game, Prisoner's Dilemma, and Implement Game)

We show that the game changes into three types by changing the parameters. Let $p(0 \leq p \leq 1)$ be the percentage paid by the user company and $1 - p$ be the percentage paid by the IT vendor. The payoffs are shown in Appendix Table 12.7.

(Condition 1) $E \geq D\Delta T$
This is the case where the effort cost exceeds the damage amount. In this case, it is reasonable to pay the damage without effort. In other words, the players make no effort to reduce problems. We call this a non-implement game.

(Condition 2) $E < D\Delta T$
The conditions under which this game becomes a prisoner's dilemma are as follows.

1. V(User company:{no effort, effort}) > V(User company:{effort, effort})
   And V(IT Vendor:{no effort, effort}) > V(IT Vendor:{effort, effort})
   Therefore,

$$1 > E/D\Delta T > p > 1 - E/D\Delta T > 0 \tag{12.5}$$

2. V(User company:{no effort, no effort}) > V(User company:{effort, no effort})
   And V(IT Vendor:{no effort, no effort}) > V(IT Vendor:{effort, no effort})
   This is consistent with (12.5).
3. V(User Company:{effort, effort}) + V(IT Vendor:{effort, effort}) > V(User Company:{no effort, no effort}) + V(IT Vendor:{no effort, no effort})
   Therefore,

$$E < D\Delta T \tag{12.6}$$

4. 2(V(User company:{effort, effort}) + V(IT Vendor:{effort, effort}) > V(User company:{effort, no effort}) + V(IT Vendor:{effort, no effort} = V(User company:{no effort, effort}) + V(IT Vendor:{no effort, effort})
   Therefore, $E < D\Delta T$. This is consistent with (12.6).

Therefore, the condition of the prisoner's dilemma is (12.5).

Table 12.7 Payoffs

		IT Vendor	
		Effort	No effort
User company	Effort	$\{2pD\Delta T - E,\ 2(1-p)D\Delta T - E\}$	$\{p\Delta T - E,\ (1-p)D\Delta T\}$
	No effort	$\{pD\Delta T,\ (1-p)D\Delta T - E\}$	$\{0,\ 0\}$

(Condition 3) $\mathbf{1 > \boldsymbol{p} \geq \boldsymbol{E/D\Delta T}}$ **or** $\mathbf{1 > 1 - \boldsymbol{p} \geq \boldsymbol{E/D\Delta T}}$

$$1 > p \geq E/D\Delta T \tag{12.7}$$

$$1 > 1 - p \geq E/D\Delta T \tag{12.8}$$

If condition (12.7) holds and (12.8) does not hold, the user company makes an effort but the IT vendor does not. In the opposite case, if condition (12.8) holds and (12.7) does not hold, the IT vendor makes an effort but the user company does not. When both (12.7) and (12.8) hold, both the user company and the IT vendor make an effort. We call this an implement game.

Reference

Axelrod, R. (1997). *The complexity of cooperation*. Princeton University Press.

Bendor, A.,& Swistak, P. (2001). The Evolution of Norms. *American Journal of Sociology, 106*(6), 1493–1545.

Chapter 13
Systemic Approach to Reliability and Safety Management Incorporating Uncertainty

Kenji Tanaka

Abstract Although reliability and safety engineering theories have previously been systematized as a related discipline, numerous accidents continue to occur in large-scale complex systems, primarily because those theories lack a system perspective. This chapter will introduce a new approach to considering the reliability and safety of real-world systems operating in diverse environments by introducing the concept of emergent failures caused by component interactions, along with a model that focuses on the gray zone (GZ) between normal and failure (or the GZ between safety and danger), which is not found in conventional theories based on two-valued logic.

Keywords Reliability theory · Emergent failure · Safety theory · Gray zone · System thinking

13.1 Introducing a Systems Perspective to Reliability and Safety Theory

Although research on reliability engineering first started around 1950 and even though the field has already been systematized as a discipline, real-world system accidents continue to occur. This is because reliability engineering theory lags behind technological developments and is not sufficiently grounded in actual phenomena (Henley & Kumamoto, 1985; Kumamoto & Henley, 1996).

A typical example of this is the misconception that a system constructed by combining highly reliable elements will always be highly reliable, which is not necessarily true. This misunderstanding is due to the lack of a system perspective. In a real-world system, low-reliability system can be introduced even if the system is assembled with high-reliability elements due to the creation of various new interactions. In such a situation, it is necessary to develop a reliability theory that focuses

K. Tanaka (✉)
University of Electro-Communications, Chōfu, Tokyo, Japan
e-mail: tanaka@is.uec.ac.jp

K. Kijima et al. (eds.), *Systems Research II*, Translational Systems Sciences 27,
https://doi.org/10.1007/978-981-16-9941-2_13

on system thinking such as emergence and hierarchy (Checkland, 1981). From this point of view, I have previously proposed the concept of "***Emergent Failure***," which incorporates emergent properties, and have advocated its importance. The contents of this proposal are introduced in Sect. 13.2.

Safety theory, which was established on a binary model of safety and danger, has pursued ensuring safety and preventing accidents by avoiding danger. However, in real-world systems, there is often a gray zone (GZ) containing potential risk factors between safety and danger within which many accidents occur. Similarly, even though the development of reliability theory based on a binary model of normal and failure has helped avoid numerous failures, many potential real-world failure factors that are hidden between normal operations and failure conditions have often been overlooked.

Among the reasons for the existence of such GZs is that the number of components in individual systems has increased enormously with the increasing scale of those systems, which means the interactions between those components have multiplied as well. Another factor is that those burgeoning integrations have created "black boxing" phenomena that make systems increasingly opaque and thus more difficult for users to understand. As a result, even though measures are taken to prevent obvious hazards and failure phenomena, the limits of human perception and prediction in uncertain situations within those GZs have not been sufficiently considered.

With these points in mind, we have proposed a new model that incorporates such GZs, and investigated the causes of many accidents based on this model. In addition, we have advocated safety and high-reliability design methods that consider uncertainty, such as a ***safety assurance design*** and a ***danger avoidance design***, as will be discussed in Sect. 13.3.

Furthermore, in today's highly automated society, the adaptive judgment and responsiveness of human beings are indispensable tools for dealing with GZs, and human–machine (H-M) systems are desired due to their high reliability. The mathematical model conditions confirming the superiority of H-M systems over complete automated systems are clarified in Sect. 13.4. Finally, Section 13.5 states our goal is to establish a ***systemic safety management***, which can only be achieved by integrating reliability and safety theory with general systems theory.

13.2 Emergent Failure Generation

13.2.1 Reliability Engineering Theory from a Systems Perspective

Reliability is never an inherent property of an object. According to the Japan Industrial Standards (JIS) definition, reliability is a value that depends on given conditions and a property that is determined by the environment in which the object is used as well as how it interacts with its surroundings. The environment, as defined here, includes not only ambient environmental conditions such as temperature and humidity but also the physical environment around the target item, its relationship with the people involved, and all other elements surrounding the target.

Taking a failure in which an electronic control device that was installed near an engine in automobile had partially melted due to heat as an example, it is clear that problems related to interaction-based phenomena could not be avoided no matter how many reliability tests had been conducted on the electronic control device itself.

At the component level, interactions increase as component numbers and integrations rise, while at the module level, between-module interactions are becoming increasingly more opaque due to black boxing. As a result, it is necessary to tackle reliability issues from one level above each hierarchy while considering such interactions. This interaction-aware approach to reliability is a new perspective based on a systems perspective that has not previously been emphasized in conventional reliability engineering theory.

13.2.2 Emergent Failures

I call the troubling phenomenon of failures occurring through the interaction of normal components **emergent failures** (Tanaka, 2014) based on their relationship to emergent properties. These are properties that cannot be explained by decomposing them into elements and are among the most important systemic properties in systems theory (Checkland, 1981). For example, the boiling point of water cannot be explained by examining the properties of its constituent atoms, hydrogen and oxygen. Instead, it requires knowledge of their arrangement structure at the molecular level.

In systems theory, we often refer to desirable properties that arise from interactions that are not found in the constituent elements. In terms of reliability, however, undesirable problems often arise due to failures that occur in combinations of normal components. We call this the emergent failure phenomenon, and it is an important consideration because pursuing failures in disassembled elements will not necessarily lead to the discovery of failures at higher levels. Therefore, to ensure that emergent failures are not overlooked, the mindset of always looking at the target at a higher level is required:

The following three patterns are typical examples of emergent failures:

Pattern A: Failures that are caused by non-contact interactions.
Pattern B: Failures that are caused by component combination incompatibilities.
Pattern C: Faulty contact failures.

Examples of these three patterns are provided below.

13.2.2.1 [Pattern A] Failures Caused by Non-contact Interactions

The electromagnetic interference caused by cell phones is a well-known non-contact interaction problem, and related malfunctions have been confirmed not only in medical equipment but also in electric wheelchairs. Because of this, a requirement to test resistance to electromagnetic waves and static electricity was added to JIS (January 2001). The following related daily life problems have also been reported.

Case 13.1 Trouble Due to Remote Control Interference
In 2007, an incident was reported in which there was a possibility of a fire due to remote control signal confusion. While operating a DVD recorder via remote control, the switch of an electric heater placed in the same room was suddenly activated. The reason was that a part of the control code governing the operation of the DVD recorder remote control coincidentally matched a part of the control code for the electric heater remote control.

In Japan's home appliance industry, each company allocates signals according to the industry's voluntary standards in order to avoid interference with the products of other companies. For this reason, there is no possibility of interference when domestic products are used in proximity with each other, but there is no such arrangement for imported products.

This abovementioned accident led to a ban by the Ministry of Economy, Trade and Industry (METI) on the domestic manufacture and foreign import of electric heaters that can be switched on via remote control. However, the incident highlights the point that potential interactions due to the simultaneous use of different products in proximity with each other are difficult to detect in laboratory tests.

Case 13.2 Automatic Brake Malfunction
An accident occurred in which a radar device installed in a car to measure the distance between it and the car immediately ahead of it received a diffuse radio wave, thus causing the car to brake suddenly by mistake and be rear-ended by the following car. This incident was caused when the outgoing radio wave from the sensor-equipped vehicle hit the side of a tanker truck driving in the next lane and bounced off, thus causing the receiving senor to misidentify the tanker as the car in front and activate the automatic brakes (Fig. 13.1).

The reliability of the radar system itself, which was devised to measure the distance between vehicles, had been confirmed via reliability tests in its design or development phase. The problem is that it was impossible for those tests to predict how the system would be affected from surrounding environments during real-world use.

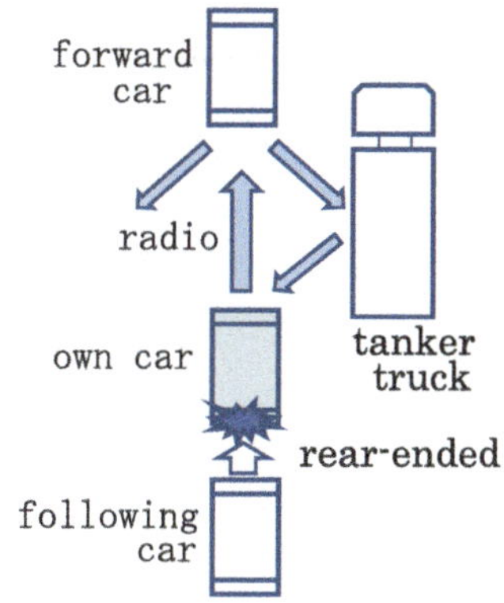

Fig. 13.1 Automatic brake malfunction

13.2.2.2 [Pattern B] Failures Caused by Component Combination Incompatibilities

Recent products have numerous components that are combined in various ways. In the incident described below, a fatality occurred due to component combination incompatibility.

Case 13.3 Ventilator Suffocation Accident
In March 2001, a suffocation accident involving an infant ventilator occurred in a hospital in Tokyo. This ventilator was designed to connect a tracheostomy tube implanted in the patient to an external oxygen cylinder via a Jackson Reese circuit to assist breathing. However, coincidentally, the exhalation outlet caliber on the patient side of the tracheostomy tube manufactured by Company A coincided with the caliber of the inner tube of the oxygen supply side in the Jackson Reese circuit manufactured by Company B. As a result, when they were connected, the oxygen supply and exhaled air from the patient collided, thereby causing flow stagnation, respiratory failure, and the patient's death via suffocation (Fig. 13.2). Here, it should be noted that both the tubing and the circuit outlets were products that conformed to Japanese domestic standards and that both parts are manufactured and sold by several companies. The inappropriate situation that led to the patient's death only occurred when the parts produced by the above two companies were coincidentally combined.

In this case, reliability tests were probably conducted on each component, but it is believed that integration tests with other components were not conducted. It goes without saying that testing at a combined level is required and that manufacturers carry out such testing at each hierarchy level for their own components. However, sufficient testing may not be carried out to ascertain what might happen when users combine their components with components manufactured by different companies. That was a blind spot.

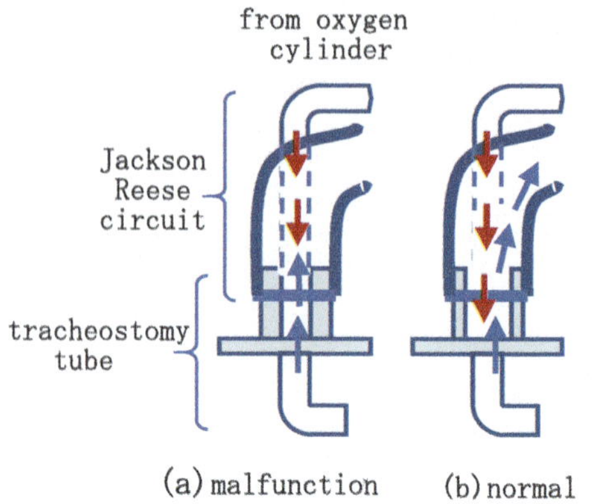

Fig. 13.2 Combination of incompatible components of ventilator

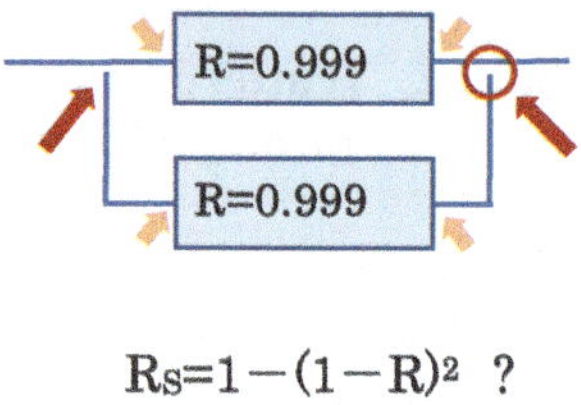

Fig. 13.3 Reliability of redundant parallelization

13.2.2.3 [Pattern C] Faulty Contact Failure

There are many cases in which contact points fail, leading to accidents. This type of failure occurs at the seam where different parts, components, and devices meet. For example, in 2007, an aircraft fire broke out at Naha Airport in Okinawa. The root cause of the fire was the falling off of a bolt designed to support the moving slats of the aircraft's wing. This case illustrates the importance of reliable contact points. Falling bolts have been responsible for other serious accidents, including a brake failure on the Tokaido Shinkansen that affected 50,000 people in May 1992.

While immense attention is paid to ensuring the high reliability of each component, there is a tendency to pay less attention to the reliability of the contact points and harness lines connecting them. This problem extends even to reliability textbooks, where, in the calculation of reliability of redundant parallelization, it is only written that the reliability of the system Rs can be calculated by $1-(1-R)^n$ when n elements with reliability R are parallelized, and there is no specific mention of contact point reliability. However, since the reliability of the contact point where the elements are put together is not equal to one, reliability will lower that using the above expression (Fig. 13.3). In the case of parallel systems where there are numerous contact points, the overall reliability can be much less than expected. In addition to structural seams, there are many trouble-prone contact points, such as contact areas between different materials and seams between management organizations.

Furthermore, reliability degradation has been occurring not only in the design process but also in maintenance work in recent years. For example, after a system has been disassembled and cleaned for maintenance, improper reassembly may result in loose or unconnected contact points, thus causing accidents. In order to avoid these problems, efforts have been made to integrate parts into structures that do not require disassembly and reconnection work.

It is also necessary to ensure contact point failure phenomena are not overlooked during the enumeration stage of failure modes and effects analysis (FMEA). During fault tree analysis (FTA), since factors are often classified and listed according to the equipment hardware aspects of reliability block diagrams, problems related to connections between equipment components are easily overlooked. For this reason, it is necessary to focus on connection functions and examine them in detail.

13.3 Gray Zone Model

13.3.1 What Is a Gray Zone?

When we think about safety issues, we usually do not think about whether something is safe or dangerous, but rather whether it is safe or not, or whether it is dangerous or not. However, these are not two sides of the same coin. The following is a typical example.

Case 13.4 Dealing with RVs with Kangaroo Bars

When recreational vehicles (RVs) equipped with front-mounted kangaroo bars (to protect them from damage during accidental collisions with wildlife in rural areas) were sold in the market, these cars collided with pedestrians, causing frequent fatal accidents. As a result, automotive insurance companies in Europe and the U.S. stopped insuring those kinds of vehicles on the grounds that their safety could not be guaranteed, even though their danger had not been proven. Japanese insurance companies, on the other hand, judged that there was no obvious problem and continued to insure them based on the theory that there was no need to reject them until the danger had been proven. It was later determined that kangaroo bars were extremely dangerous to pedestrians and, in many cases, caused internal ruptures during collisions.

The Western response, which placed the GZ outside of safety, was of the safety/non-safety paradigm, while the Japanese response, which placed the GZ outside of danger, was based on the non-danger/danger paradigm. Hence, it is clear that determining how to deal with the gray area that exists between safety and danger, which is often difficult to estimate, is an important point of consideration when constructing a theory of reliability and safety in today's diverse society.

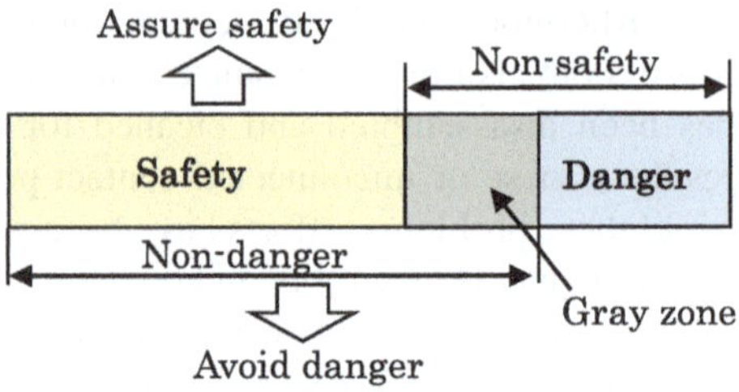

Fig. 13.4 Existence of gray zone

13.3.2 Safety Assurance Design and Danger Avoidance Design

Based on the GZ model, we previously proposed two safety design methods (Tanaka, 2011). One is the ***safety assurance design***, which aims to avoid the GZ to ensure the product can be used safely, and the other is the ***danger avoidance design***, which aims to allow the product to be used in the GZ but avoid specific hazards (cf. Fig. 13.4).

A typical example of a safety assurance design is a fool-proof design. For example, even if you try to open the lid of a washing machine during its spin cycle, you find that the lid is locked and will not open. This feature blocks potentially dangerous actions and uncertain usage in the GZ and functions safely only when used correctly. On the other hand, an example of a risk-averse design is a structure that can be loosened but not removed completely, such as fasteners in places where a falling bolt might cause a fatal injury.

These two design categories are also useful in user manuals and warning labels. User manuals are mainly intended to provide information on the correct usage and operation of the product and thus promote safety and GZ avoidance. In contrast, warning labels are used to caution against dangerous misuse and are not intended to avoid the GZ. Therefore, accidents in the GZ sometimes occur despite the hazard warning signs, such as the one described below.

Case 13.5 Accident in Hyperbaric Oxygen Therapy
In hyperbaric oxygen therapy, the patient enters a dome where he or she is subjected to high atmospheric pressure. In this case, a patient wearing pajamas containing 5% synthetic fibers entered a hyperbaric oxygen therapy dome and suffered burns, despite the warning label stating that "Garments made of synthetic fibers are dangerous."

From the warning label, we can readily understand that it would be dangerous to enter the hyperbaric oxygen therapy dome wearing clothing articles consisting of 100% synthetic fibers, but it is difficult to understand that garments with just 5% synthetic fibers would be dangerous as well. This accident could have been prevented if there had been a label instructing users to verify the safety of their garments, such as "Wear 100% cotton garments only," instead of a caution label on the dangerous side.

13.3.3 Toward Safety Acquisition Considering the Gray Zone

It is necessary to properly use safety assurance and danger avoidance according to its intended purpose. Safety assurance type of design or operation indication is effective in the following two case types:

1. When an erroneous operation will have a significant impact on life.
2. When the user has little prior knowledge about the product.

In the first case type, it is necessary to ensure that users are informed of the correct way to use the system. Recently, home appliances have become more sophisticated, and there are increasing numbers of products whose internal structures are not readily understandable. Accordingly, it is desirable to adopt designs that naturally prompt users to operate the products correctly even if they do not read the instruction manual, such as by incorporating affordances or adopting fool-proof structures.

On the other hand, there are cases in which warning labels and caution signs can be effective, such as the following:

1. Avoidance of unintentional misuse that can be easily caused by experience and knowledge.
2. Discouraging intentional misuse.

The "safe operation" indications in user manuals or regulations can be highly effective, but there are sometimes problems with excessive use (Table 13.1). In fact, safety assurance type expressions tend to impose strong restrictions on usage. In Case 13.5, there may be cases when garments can be worn safely even if they are not 100% cotton, but this option is forbidden. In general, if usage restrictions are too strict, there is a possibility that users will ignore the guidance and deviate to the GZ area. On the other hand, danger avoidance types have GZ risks, but their use is more flexible and allows more freedom.

If we can reduce a GZ size by collecting information, we can eliminate accidents inside the GZ. However, in many cases, it is difficult to eliminate the GZ itself. Rather, it is expected that users will learn how to use the system GZ wisely, which will lead to enhanced safety and security. For this purpose, it is necessary to minimize safety assurance types and increase the number of danger avoidance types so that users themselves can develop appropriate adaptive response skills (Fig. 13.5).

Table 13.1 Safe assurance and danger avoidance types

Design/indication	Advantages	Disadvantages
Safety assurance types	Avoids GZ use Can be used with confidence	Strict usage constraints Deviations
Danger avoidance types	Flexible GZ use	Some GZ risk

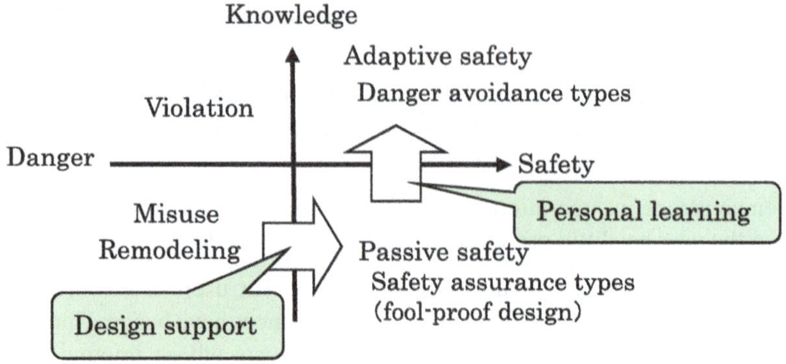

Fig. 13.5 From passive to adaptive safety

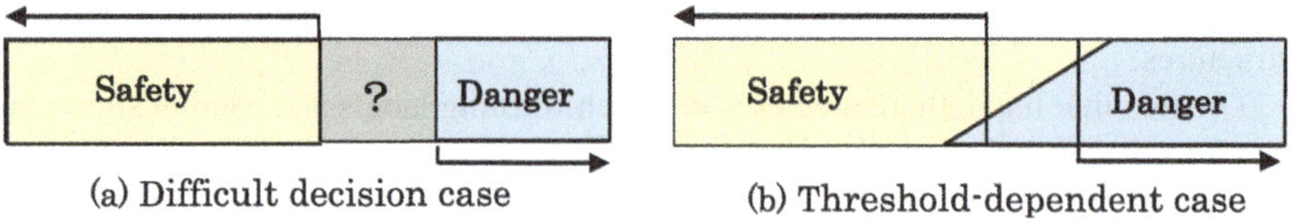

Fig. 13.6 Two gray zone case types

13.3.4 Threshold-Dependent Cases

Analyzing a large number of accidents has made me realize that GZ cases could be divided into two types (Fig. 13.6).

The first type consists of GZs where it is impossible to determine whether the situation is safe or dangerous due to a lack of information. We call these ***difficult decision cases***, as has been discussed up to Sect. 13.3.3. The second is a new type involving cases where the GZ is created due to differences in the way the danger threshold is set. For example, in cases involving judging radiation danger levels, nuclear proponents and local residents will commonly have different judgment thresholds and disagree with each other. This is because nuclear power proponents think in terms of danger and non-danger, while residents think in terms of safety and unsafety. We call these ***threshold-dependent cases***.

Rules and regulations may often be subject to misinterpretation between those who enact them and those who follow them, and trouble may occur as a result. Following the Great East Japan Earthquake that occurred on March 11, 2011, rumors about radioactive contamination of vegetables arose, but it was unclear whether the standard was on the safe side or the dangerous side, so anxiety spread.

Case 13.6 Allowable Amounts of Radioactive Materials in Vegetables
The provisional standard value for radioactive iodine in spinach and other vegetables was set at 2000 becquerels per kilogram. However, if one were to eat 100 g of the

standard level of spinach every day for 1 year, the annual radiation exposure would be

$$2000 \times 100/1000 \times 365 \text{ days} \times 0.000016 = 1.17 \text{ millisieverts}$$

This level is considered safe and is less than the provisional standard of 5 millisieverts per year, which is considered to be a safe level of exposure from food and drink.

According to a 1992 report by the International Commission on Radiological Protection (ICRP), even a standard value of 10 millisieverts is "not inappropriate," so the 2000-becquerel level can be considered to be on the safe side. However, the Japanese public, informed by various media reports, has fallen into the trap of believing that anything above the 2000-becquerel threshold is dangerous.

In threshold-dependent cases, it is desirable to eliminate the GZ by aligning the safety boundary with the danger boundary. Even if this is difficult, it is at least necessary to understand how each position can be interpreted. In other words, since GZ elimination is difficult, it is very important to understand whether a rule is on the dangerous side (and must be followed absolutely) or whether it represents an ideal boundary (to ensure safety). Particularly since interpreting it in the opposite sense could have serious consequences if an accident occurs.

13.4 Automated Human–Machine Cooperative Systems

In recent years, the development of automation technology has been progressing, and self-driving cars have even begun to operate on public streets. Since highly accurate monitoring technology is required in an automated society, we have already formulated this problem with a logical model and analyzed it mathematically. The main results obtained are introduced below.

13.4.1 Non-Homogeneous Safety Monitoring System

In previous studies in monitoring systems to guarantee safety, two primary failure problems have been considered: ***failed dangerous (FD)*** and ***failed safe (FS)*** failures (Henley & Kumamoto, 1985). FD failure refers to a sensor that fails to generate an alarm signal when the monitored object is in danger, while FS failure refers to a sensor that generates an alarm signal when the object is actually safe. Safety monitoring systems have been developed to prevent both FD and FS failures of individual sensors by the use of a method that generates an alarm based on information from several sensors.

However, traditional two-state models do not consider GZs in which neither safety nor danger can be confirmed by the sensors, and the lack of information about danger often leads to erroneous safety signals. This means that a safety monitoring system must be able to distinguish between the lack of information about danger and the presence of information about safety in the monitored object.

We introduced two kinds of sensors to make this distinction: ***safety-presentation (SP)*** and ***fault-warning (FW)*** sensors (Tanaka, 1995). An SP-sensor generates a safety signal only when it confirms that the monitored object is safe. However, when an SP-sensor does not generate a safety signal, it does not necessarily mean that the object is in danger. The safety signal absence may be caused by a lack of relevant information. This means that an SP-sensor is a safety assurance type of sensor.

Similarly, the FW-sensor is a danger avoidance type of sensor and generates a danger signal only when it confirms that the object is in danger. Both sensor types are easily produced by setting different thresholds, T_{SP} and T_{FW}, on otherwise equivalent sensor devices (cf. Fig. 13.8).

Next, we focus on two-sensor systems consisting of SP- and FW-sensor pairs, which monitor the same characteristic of the monitored system independently of one another. Such two-sensor systems, which are referred to as ***non-homogeneous safety monitoring systems*** (NHSMS), can be expected to decrease the occurrence of both FD and FS failures in GZs.

We show that an NHSMS can identify a GZ based on the outputs of the two sensors. Figure 13.7 shows four output patterns of an NHSMS where the safe state identified by the SP-sensor and the danger state identified by the FW-sensor are denoted as S and D, respectively. In this figure, X indicates that neither sensor can detect a state.

Pattern (a) is a standard arrangement seen when the monitored object is safe according to the safety signal generated by the SP-sensor and no danger signal from the FW-sensor ($S \cap X$). In contrast, Pattern (b) is the standard arrangement that indicates danger ($X \cap D$). In Pattern (c), both the safety and danger signals are absent ($X \cap X$), thus indicating a GZ in which neither safety nor danger can be confirmed. In Pattern (d), both safety and danger signals are present, which represents a situation in which the states of the monitored object cannot be determined because one or both sensors have failed ($S \cap D = \phi$).

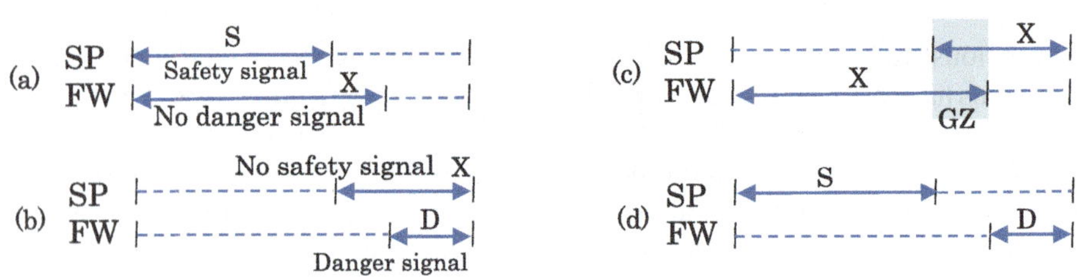

Fig. 13.7 NHSMS output patterns

In NHSMS, GZ-related judgments are crucial. In the case of an automatic NHSMS, GZ judgment rules must be determined in advance. Hence, two system types of NHSMS are considered, one that interprets the GZ as safe and the other that interprets it as dangerous. By modeling and analyzing those systems, Tanaka (1995) derived conditions under which an automatic NHSMS consisting of an SP- and FW-sensor setup is more reliable than systems with two SP-sensors and two FW-sensors. Another method to judge the GZ is to inspect the state in detail and to judge from added information by human.

13.4.2 Human–Machine Cooperative Monitoring System

In H-M system GZs, operators or experts are expected to judge actual system states based on information obtained from detailed inspections or related sensors outputs, after which additional inspection activities by relevant humans may impose additional costs and introduce inspection errors. When the GZ is identified by Pattern (c) in Fig. 13.7, inspection activities should be performed provided that time is sufficient and the cost and potential for human error are suitably small.

As mentioned above, the following two monitoring strategies can be used for determining the state of the monitored object in the GZ (Fig. 13.8).

A. ***Automatic monitoring*** in which either *S* or *D* is defined prior to the operation:
   When the GZ is considered to be a danger zone (GZ = D), the system is referred to as a ***sensitive (S-) system.*** In contrast, when the GZ is considered to be a safe zone (GZ = S), the system is referred to as a ***blunt (B-) system***. Consequently, those monitoring systems automatically judge the state of the monitored object.
B. ***H–M cooperative monitoring*** in which the state of the monitored object is judged based on information from detailed inspections by operators or inspectors:

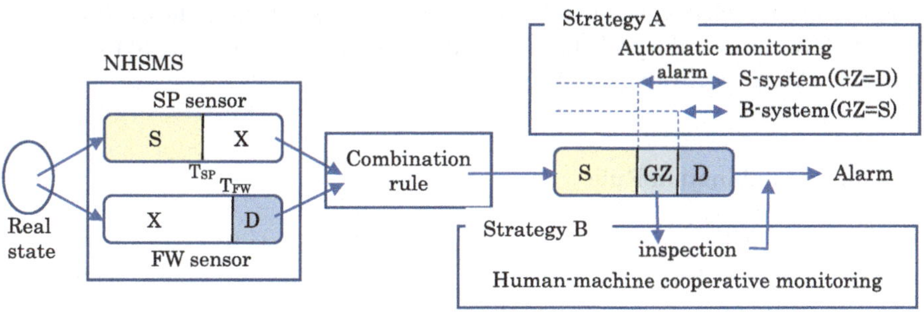

Fig. 13.8 Two NHSMS model strategies

For Patterns (a) and (b), the system automatically judges the states, while for Pattern (c), the system judges states after cooperative H-M activities. For Pattern (d), either automatic or human judgment is selected.

Here, it is necessary to clarify the criteria for selecting the optimal type among a H-M cooperative monitoring system with inspection, and two types of automatic NHSMS without human inspections. In each of these types, ordinary monitoring is performed automatically, and the only difference is in the way in which the GZ state is judged. In an H-M cooperative monitoring system, the judgment is made by human operators or experts.

An optimal design is one that minimizes losses due to FD/FS failures and reduces inspection costs and errors. In order to obtain conditions under which an H-M cooperative system is more effective than automatic monitoring systems, we employed a method based on the Dempster-Shafer theory (DST) (Shafer, 1976) to deal with problem uncertainties (Tanaka & Klir, 1996, 1999).

13.4.3 Mathematical Approach to Optimal Monitoring Strategy

Since the DST, which is also known as the theory of evidence, is a powerful theory for dealing with uncertainty (Shafer, 1976), we adopted it in our pursuit of an optimal monitoring strategy. For a detailed mathematical description of our analysis, please refer to (Tanaka & Klir, 1999). In this section, only the results are presented. The two prominent characteristics of the theory are domain assigning a probability and the use of a combination rule.

13.4.3.1 Domain Assigning Probability

In DTS, X has a similar meaning to a universal set under interest and is usually referred to as the frame of discernment. The basic probability assignment (BPA) is defined as a function m: $\{S,D,X\} \rightarrow [0,1]$. X includes a safety domain S or a danger domain D as its subsets such that $S \cap D = \phi$. When m_{SP} and m_{FW} show BPAs of the SP- and FW-sensors, the probabilities of no sensor signal are expressed by $m_{\mathrm{SP}}(X)$ and $m_{\mathrm{FW}}(X)$, respectively (Fig. 13.9).

13.4.3.2 Combination Rule

An important DST characteristic is the way in which multiple BPAs are combined to obtain a joint BPA. This is referred to as a combination rule. As several combination rules have been proposed in the literature, the optimal rule for our purpose must be selected in order to combine the outputs of the two sensors.

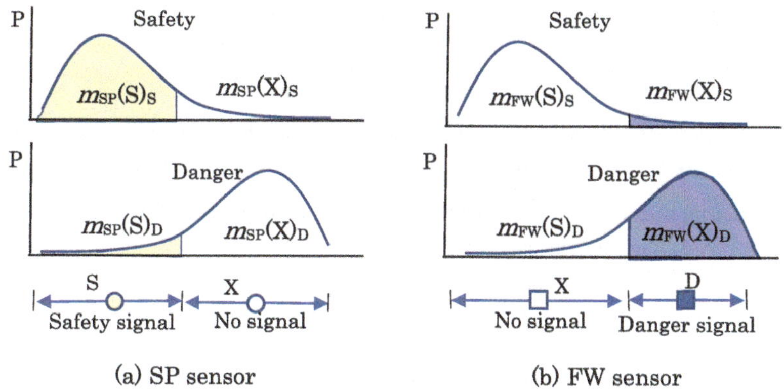

Fig. 13.9 BPAs of SP-and FW-sensors

The standard first step of combination is expressed by

$$q(S) = m_{\mathrm{SP}}(S)m_{\mathrm{FW}}(X), \quad q(D) = m_{\mathrm{SP}}(X)m_{\mathrm{FW}}(D),$$
$$q\,(\mathrm{GZ}) = m_{\mathrm{SP}}(X)m_{\mathrm{FW}}(X), \quad q\,(\phi) = m_{\mathrm{SP}}(S)m_{\mathrm{FW}}(D).$$

Next, a joint BPA m is obtained by normalizing function q to satisfy the requirement that $m(\phi) = 0$. In the proposed model, ϕ corresponds to a case in which both a safety signal and a danger signal are generated simultaneously due to the physical failure of one or both sensors. At the design stage of the safety monitoring system, we need to determine how to deal with this case.

Yager's rule (Yager, 1987) is given by

$$m(C) = q(C), \quad (C = S, D) \quad m\,(\mathrm{GZ}) = q\,(\mathrm{GZ}) + q\,(\phi)$$

and regards contradictions as unknown states. Inagaki (1991) referred to the rule given by

$$m(C) = \{1 - q(X)\} / \{1 - q(X) - q\,(\phi)\} * q(C), (C = S, D)$$
$$m\,(\mathrm{GZ}) = q\,(\mathrm{GZ})$$

as the extra rule. The extra rule maintains only $q(\mathrm{GZ})$ invariably.

In our model, selecting a combination rule means determining how to cope with conflicting information from sensor outputs. Yager's rule dictates that the conflict pattern is regarded as an uncertainty domain. However, the extra rule dictates that the conflict pattern is to be judged as safe or dangerous based on historical data.

13.4.3.3 Assumption

We assume that

[A1] two sensors have identical and independent distributions; and
[A2] the distribution of each sensor is symmetric with respect to the threshold.

[A2] means that the correct response rate of an SP-sensor for safety is equal to that of an FW-sensor for danger; that is,

$$m_{\mathrm{SP};S}(S) = m_{\mathrm{FW};D}(D) = a$$

It also implies that the wrong response rate of an SP-sensor for danger is equal to that of an FW-sensor for safety; that is,

$$m_{\mathrm{SP};D}(S) = m_{\mathrm{FW};S}(D) = c.$$

In addition, assume that

$$[\mathrm{A3}]\, 1 - a - c = \theta > 0, \quad [\mathrm{A4}]\, a \geq c.$$

If θ is zero, an SP-sensor cannot be distinguished from an FW-sensor. Hence, [A3] is needed. [A4] means that the correct response rate is greater than the wrong response rate.

13.4.3.4 Optimal Monitoring System

The total expected loss due to FD/FS failures is used as an evaluation function to obtain the optimal system (Klir & Wierman, 1998).

Strategy A. Optimal Automatic Monitoring

We focus on the automatic monitoring system type and compare a sensitive-type system with a blunt-type system. In addition, the optimal combination rule is determined. To perform this analysis, the following terms are introduced:

$$K = \pi C_{\mathrm{b}}, \quad L = (1 - \pi)\, C_{\mathrm{a}} \quad \text{and} \quad \alpha = (1 - c)\, a / \,(1 - a)\, c$$

where $\pi = p(D)$ is the demand probability, C_{a} is the loss caused by one FS failure, and C_{b} is the loss caused by one FD failure.

High α values reveal that the sensor is reliable. Based on assumption [A4], α is greater than one. The optimal monitoring system type and the optimal combination rule are specified in the following proposition.

Proposition 13.1 In automatic monitoring systems:

1. When $L \leq K$, the optimal type is an *S-system* and the optimal rule is

$$\begin{cases} \text{Yager's rule if } 1 \leq \alpha \leq K/L \\ \text{extra rule} \quad \text{if } K/L \leq \alpha \end{cases}.$$

2. When $L \geq K$, the optimal type is a *B-system* and the optimal rule is

$$\begin{cases} \text{Yager's rule if } 1 \leq a \leq L/K \\ \text{extra rule} \quad \text{if } L/K \leq \alpha \end{cases}.$$

Proposition 13.1 reveals that the optimal monitoring type depends solely on the ratio of K and L, which are relevant to the monitored system, and that it does not depend on the reliability of the sensors.

Strategy B. Human–Machine Cooperative Monitoring

Next, consider an H-M cooperative monitoring system in which the object is inspected when the system output is GZ. As noted previously, when neither the SP- nor the FW-sensors generate a signal in the NHSMS, the object is considered to be in the GZ. In this case, the operators must determine the state of the object based on other information or detailed inspection data, or they must ask inspection experts. Although operators or inspection experts can adaptively judge the state of the object from additional information obtained via inspection, the potential for data loss from incomplete inspections must be considered when deciding whether to adopt the H-M cooperative monitoring system. Incomplete inspections can result from two causes:

1. inspection fails to detect the actual danger, or,
2. damage occurs before the inspection is completed.

Let R, which is calculated by the mean value of the probability of incomplete inspections for predictable failure modes, denote the total probability of incomplete inspections. When the object is in danger, the expected value of the damage due to incomplete inspection RC_{b} should be added. If the inspection is always completed, R should be set to zero. Let C_{I} denote the inspection cost. When the inspection is performed, the inspection cost C_{I} is incurred even if the object is safe. By using these values, the expected loss for the H-M cooperative monitoring is obtained.

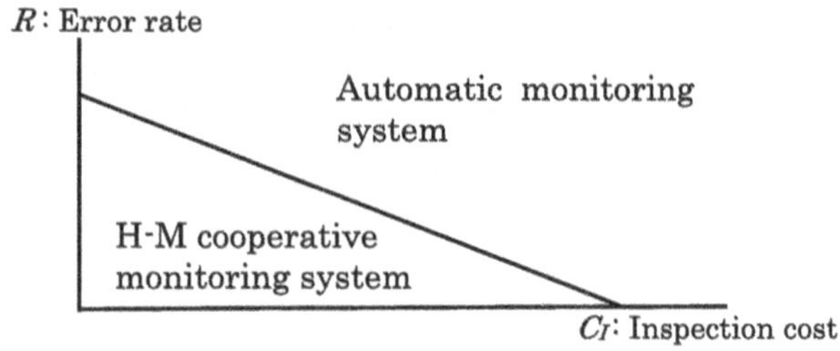

Fig. 13.10 Optimal monitoring system

The next proposition shows the optimal combination rule for an H-M cooperative monitoring system.

Proposition 13.2 In H-M cooperative monitoring system, the optimal rule is

$$\begin{cases} \text{Yager's rule} & \text{if } (K+L)/(C_{\mathrm{I}}+KR) \geq 1+\alpha \\ \text{extra rule} & \text{if } (K+L)/(C_{\mathrm{I}}+KR) < 1+\alpha \end{cases}.$$

The next Theorem 13.1 shows the optimal monitoring type selected between the optimal automatic monitoring derived in Proposition 13.1 and H-M cooperative monitoring with inspections.

Theorem 13.1 The H-M cooperative monitoring system is more effective than an automatic monitoring system if and only if $\min(K, L) - RK > C_{\mathrm{I}}$.

Theorem 13.1 shows a strict intuitive inference that NHSMS with inspection is more desirable than automatic monitoring when the inspection cost and inspection error rate are minimized (Fig. 13.10).

13.5 Toward Establishing a Systemic Safety Management

Although I initially thought that eliminating GZs would be desirable, further study indicated that GZs could be used to enhance risk awareness and response capabilities in difficult decision cases. Furthermore, while it is better to avoid obvious hazards by design, by keeping a GZ, a system can be used in a variety of ways, particularly for helping people improve their ability to consciously catch risks and develop adaptive response skills.

Thus, my current goal is to take a systemic approach to reliability and safety engineering theories from a management perspective, including the improvement of human response capability in the gray zone, and ultimately establish a "***Systemic Safety Management.***"

References

Checkland, P. (1981). *Systems thinking, systems practice*. Wiley.

Henley, E. J., & Kumamoto, H. (1985). *Designing for reliability and safety control*. Prentice–Hall.

Inagaki, T. (1991). Interdependence between safety control policy and multiple-sensor schemes via Dempster–Shafer theory. *IEEE Transactions on Reliability, R40-2*, 182–188.

Klir, G. J., & Wierman, M. A. (1998). *Uncertainty-based information: Elements of generalized information theory*. Physica Verlag/Springer.

Kumamoto, H., & Henley, E. J. (1996). *Probabilistic risk assessment and management for engineers and scientists*. IEEE Press.

Shafer, G. (1976). *A mathematical theory of evidence*. Princeton University Press.

Tanaka, K. (1995). Reliable design of safety monitoring system accounting for the uncertainty zone. *Reliability Engineering and System Safety, 48-2*, 75–84.

Tanaka, K. (2011). Consideration on how to properly use safety assurance and danger avoidance-instruction manual and warning display. In *41th reliability & maintenance symposium* (pp. 373–378). JUSE. (in Japanese).

Tanaka, K. (2014). *System reliability and safety*. Asakura Publishing. (in Japanese). (Received the Nikkei QC Literature Prize in JAPAN).

Tanaka, K., & Klir, G. J. (1996). A design method for incorporating human judgement into monitoring systems. In T. Sheridan (Ed.), *Analysis design and evaluation of man–machine systems 1995* (pp. 717–722). Pergamon.

Tanaka, K., & Klir, G. J. (1999). A design condition for incorporating human judgement into monitoring systems. *Reliability Engineering and System Safety, 65*, 251–258.

Yager, R. R. (1987). On the Dempster–Shafer framework and new combination rules. *Information Sciences, 41*, 93–137.

The manufacturer's authorised representative in the EU is Springer Nature Customer Service Centre GmbH, Europaplatz 3, 69115 Heidelberg, Germany. If you have any concerns regarding our products, please contact ProductSafety@springernature.com

Printed and bound by CPI Group (UK) Ltd, Croydon, CR0 4YY
30/09/2026
02225329-0007